T0314096

Beijing Garbage

Consumption and Sustainability in Asia

Asia is the primary site of production of a myriad of commodities that circulate the globe. From cars and computer chips to brand clothing, material objects manufactured across Asia have become indispensable to people's lives in most cultural contexts. This mega production generates huge amounts of waste and pollution that threaten the health and lifestyle of many Asians. Yet, Asia is not only a site of production, but also one of the most rapidly growing consumer markets.

This series focuses on consumption – the engine propelling Asia onto the world economic stage – and its implications, from practices and ideologies to environmental sustainability, both globally and on the region itself. The series explores the interplay between the state, market economy, technologies, and everyday life, all of which have become defining facets of contemporary Asian culture. Shifts in consumption that have taken place across Asia since the 1950s have had a deep impact on new and emerging informal economies of material care, revealing previously invisible sites of innovation, resistance and co-option. The series will bring together studies by historians, anthropologists, geographers, and political scientists that systematically document and conceptualize Asia's engagement with consumption and sustainability in the global environment.

Series Editor
Katarzyna Cwiertka, Leiden University

Editorial Board
Nir Avieli, Ben Gurion University of the Negev, Beer-Sheva
Assa Doron, Australian National University

Beijing Garbage

A City Besieged by Waste

Stefan Landsberger

Amsterdam University Press

The research for this publication took place under the auspices of the Garbage Matters: a Comparative History of Waste in East Asia Project, funded by the Netherlands Organization for Scientific Research (NWO), project number 277-53-006.

Cover illustration: Collected paper and cardboard waiting to be moved to the next higher collection point for further processing, Haidian District, Beijing
Author's photograph, April 2017

Cover design: Coördesign, Leiden
Typesetting: Crius Group, Hulshout

ISBN   978 94 6372 030 4
e-ISBN   978 90 4854 287 1 (pdf)
DOI   10.5117/9789463720304
NUR   740

Printed by CPI Group (UK) Ltd, Croydon CR0 4YY

Table of Contents

List of Illustrations

Introduction

Solid waste has become an increasingly difficult problem to deal with worldwide, particularly in the urban areas that cover an ever-growing percentage of the globe. People now buy more things, because the products they have acquired break down more quickly and easily as a result of planned or built-in obsolescence (Lucas, 2002; Clapp, 2002; O'Brien, 2013). Repairing broken goods is often more expensive and less convenient than simply buying a new product, even in nations where repairing work used to thrive (McCollough, 2007; McCollough, 2012). People buy more because they can – due to increased incomes, more leisure time, and the explosive growth of places of purchase (Featherstone, 1993; McCollough, 2007; Cooper, 2010; McCollough, 2012). Owning things has come to signify status, which used to be derived from one's work or social position. People buy more simply because others do; they buy newer versions of products that have must-have functions that are lacking in earlier versions. People desire new products because recommissioned or refurbished articles carry the stigma of being broken or soiled (Lucas, 2002). Products distinguish themselves through packaging. As Jennifer Clapp points out, 'goods are packaged to enhance their ability to travel long distances, to give them uniformity of size for efficient distribution, to keep them sanitary, to increase their convenience of use, and to make them more appealing' (2002: 162-163). Packaging is 'inherently linked with the ritual of shopping, which stimulates the desire to buy and facilitates and enhances the endless loop of consumption' (Machotka and Cwiertka, 2016: 32). To quote from a report by McKinsey & Company and Ocean Conservancy, '[S]pecifically, plastic packaging is increasingly used to promote food safety and preserve freshness and quality as products move over greater distances and have longer shelf-life requirements. Also, in an effort to cater to lower-income consumers, companies are shrinking product-distribution sizes, creating more units of packaging per gram of product' (2015: 33). Ideally, all of these discarded (packaging) materials can be recycled or recommissioned, but despite many pious pronouncements and more or less idealistic international agreements and conventions, this seems to be an unattainable goal.

There are various ways to deal with the ever-growing amounts of solid waste, to avoid creating a situation where humankind is swallowed by it. Many if not most nations have followed a comparable trajectory of waste disposal management. Traditionally, landfills have been used to remove waste from sight by burying it. However, the volume of discarded matter has grown to such proportions that landfills are no longer adequate. With (urban) populations expanding, landfills threaten to take up too much of the space that is required for other purposes, such as construction. Moreover, landfills have moved beyond the simple burying of waste materials. Over time, the nature of waste materials has changed, their quantity has grown, and the regimes governing their disposal have become more complex and costly, factoring in both hygiene risks and social demands. Landfill sites need special preparations and facilities to avoid secondary pollution in the form of potentially toxic materials that could enter the soil and contaminate ground water; the pervasive smell of rot and corruption needs to be masked, and so on. Another tried and proven method of removing waste has been recycling, if and when materials or parts of products can be reused again. Recycling tends to be labour-intensive and dirty work, and is often associated with people who have been rejected by society or who have no other way to survive (Drackner, 2005; Yates, 2006). Most recently, incineration or the burning of waste on an industrial scale, often producing marketable side products like electricity or warmth, has received strong political support and as a result strong development worldwide. However, it seems that none of these approaches will be able to solve the problem by themselves, without decreasing (over-)consumption by us, the people.

Theorizing waste and consumer culture

Over time, waste has been defined and theorized in many ways by scholars, who are mainly from Western, industrialized countries. Among the plethora of definitions and theories, Drackner has created a useful catch-all typology by arguing that waste is '*something* that is discarded by *someone*', implicating its uselessness (Drackner, 2005: 176, italics in original). Whatever the definition applied in whichever culture, dirt, waste, or non-matter needs to be eliminated in a positive effort to organize the environment (O'Brien, 1999). Mary Douglas studied ritual pollution and uncleanness, and famously saw waste, or dirt, as matter out of place; more recently, Tim Cooper considered it abject (Douglas, 1966: 36; Cooper, 2010). Be it called dirt, waste, or trash, the societies in which it appears want to get rid of it. At the same time, waste appears in dynamic categories and guises that express different values (Moore, 2012). For example, one person's waste can serve as another's livelihood (Drackner, 2005: 176). This opens avenues for contestation about waste's final displacement, a topic that has now become particularly salient. But waste is more than merely a problematic object that raises questions about its final destination; it can also be seen as something symbolically positive: a marker representing unproductive privilege, industrial efficiency, and wealth (Varul, 2006).

Pre-industrialized, proto-industrialized, or early industrialized societies were not plagued by questions of where to make non-matter end up, simply because there was not a lot of it to deal with; the small amounts of dirt, waste, or matter-out-of-place that were created were disappeared in ritual or other ways. Many if not most of these societies were characterized by a culture of scarcity and maintained an ethos of frugality (Dikötter, 2006; Van Dam and Jonker, 2017; Cwiertka and Machotka, 2018). In these frugal societies, goods were used, re-used, or re-appropriated into use until what was left had no use left. Moreover, the final manifestations of these goods tended to be of organic composition, allowing for them to disintegrate and disappear with hardly any mark or trace. As societies developed over time, their economies expanded and urbanization proceeded apace. In the eighteenth century, the amount of waste produced by households increased and questions were raised about its disposal: the increase in waste in the built-up urban environment can be partly attributed to the disappearance of the open kitchen fire, where previously most of it had been burned as fuel (Lucas, 2002; Drackner, 2005; Reno, 2014). Household waste was considered 'inefficient and arising through improper [domestic] management' (Lucas, 2002: 6) – in other words, caused by people or consumers. As populations and waste ballooned, concerns about the connection between waste and the spread of diseases proliferated. Waste thus came to be seen as unhygienic, unhealthy, and without value.

The link between processes of urbanization and the emergence of waste-as-a-problem is highly significant. Urban centres grew rapidly as a result of migration patterns (Drackner, 2005). Members of rural communities were used to composting what could not be burned, and more importantly had the space to reserve plots for this purpose; composted waste was applied to the nearby fields as a fertilizer or soil enricher. The emerging often densely populated urban communities, on the other hand, lacked spaces for proper waste disposal, and their inhabitants' knowledge and practice of proper disposal practices had dissipated, leading to an increase in perceived threats to and concerns about public hygiene and health. As waste became more prevalent and was seen as actually endangering society, it became an object that deserved more and more study. Consequently, the studies that have emerged over time as a result of this attention focus more on questions of waste management, particularly the role of the waste producer, i.e., the consumer, and less at the nature of the waste itself. Most energy has been devoted to probing the problems related to consumer waste, which is often seen as more problematic than industrial waste, even though the latter was and continues to be produced in much greater amounts and with more profound and detrimental effects on the environment (Yates, 2006; Gille, 2010; O'Brien, 2013).

Since the first identification of the consumer-as-waste-producer in the eighteenth century, each subsequent generation of waste producers has been considered more wasteful than the preceding ones. It is only logical, then, that their behaviour was seen as resulting in ever growing amounts of disposed matter. This wasteful behaviour and the waste itself have been explained as the result of the arrival of consumer culture, and, more negatively still, the 'throwaway society' (Debord, 1994 [1967]; Lefebvre, 1995 [1962]; Baudrillard, 1998 [1970]; Featherstone, 1998 [1991]; Lucas, 2002; McCollough, 2012; O'Brien, 2013; Hellmann and Luedicke, 2018). Most theories about consumer culture trace its emergence to the end of World War II, and more specifically to the 1960s, when the concept of the 'throwaway society' also emerged (Lucas, 2002); other theorists have traced its roots to 1890-1920, when mass production and mass distribution brought more products and services to ever larger numbers of people (Strasser, 2003: 379). Some scholars deplore this turn towards consumption (Baudrillard, 1998 [1970]), considering the replacement of what was valued as durable with the shallow, new, and disposable to be a debasement of society. Others see the process as a 'profound transformation of society, with the consumer society (production organized for the market) having taken over from blind production or production for production's sake' (Lefebvre, 1995 [1962]: 196) leading to a situation where 'the manufacturers of consumer goods do all they can to manufacture consumers' (Lefebvre, 2002 [1961]: 10). Mike Featherstone (1998: 13) helpfully summarizes these scholarly views into three perspectives that move from the consumed goods themselves, to their external value, to what they generate. The first touches upon the expansion of capitalist commodity production, which has given rise to the vast expansion of material culture through the increase of consumer goods and sites to purchase and consume. The second perspective focuses on how people use goods to create social bonds or distinctions. The third perspective deals with the emotional pleasures of consumption – as Tim Cooper describes it, the culture of consumption driven by advertising, disposability, and the supermarket (2010: 1118).

Many analyses of consumer society seem to be tinged with a sense of loss and nostalgia, permeated by a yearning for days gone by when consumer goods were acquired for their use value instead of their exchange value or the status they bestowed on the owner; they wax poetic about a time when

a consumer good would last a lifetime and could be passed on to the next
generation still in working order. They are also vaguely ideological and
conservative, deploring the disappearance of the time when the imagined
community to which the authors feel they belong was smaller and less
influenced or manipulated by external or foreign (mainly American) influences (Anderson, 1991; Debord, 1994 [1967]; Lefebvre, 1995 [1992]; Baudrillard,
1998 [1970]; Lefebvre, 2002 [1961]; Kaplan, 2012). These moral critiques of
escalating demand, high product turnover, and built-in obsolescence in a
society increasingly looking for leisure and convenience, are combined with
sociological analyses of economic and cultural changes relating to levels
of affluence, patterns of taste, and industrial innovation. This combination of theoretical perspectives results in the view that there is something
particularly wasteful about contemporary society; that consumers are
'uniquely profligate, ignorant, disdainful of their consumption behaviour'
compared to preceding generations (O'Brien, 2013: 20). In the opinion of
these theorists, wastefulness has evolved into a cultural force. Indeed, as
Hélène Cherrier puts it, consumerism points to the incessant acquisition
of 'new, modish, faddish or fashionable, always improved and improving'
goods; it nurtures an ideology of newness and creates a space where the
old, the past, and the worn-out have no place (2010: 260).

However, careful historical studies of consumer behaviour over longer
periods of time have demonstrated that the amounts of waste that present-
day consumers discard are not necessarily larger than those of their prede-
cessors (O'Brien, 2013). The composition of present-day waste certainly has
changed: ashes from the kitchen fire, for example, have largely disappeared
from consumer garbage, and their place has been taken by other types of
disposed matter. Moreover, reusable waste (packaging) has given way to
disposable packaging, largely as a result of greater attention to (personal)
hygiene (Clapp, 2002). Buying new is considered to be clean (Lucas, 2002:
12), resulting in an increase in packaging. Moreover, consumer goods have
come to be produced from less easily disposable materials and resources, and
their disposal often has more lasting, seriously polluting, and toxic effects
on the environment. The principle of built-in obsolescence has created vast
graveyards of broken-down and discarded goods and gadgets, the latter of
which Baudrillard (1998 [1970]) considered to be especially exemplary of
consumer culture: a product that was wasteful and shallow, for which no
further employ could be found.

Other theorists see the emergence of consumer society as the logical result
of the fact that the nature of waste itself has been transformed. Waste has
become a commodity, part of a scheme of production and consumption, or

even just another raw material or resource. As such, it is no longer merely disposable, but has itself acquired a consumption value (O'Brien, 1999: 277, 281, 282). This has created the phenomenon of waste regimes – social and political constellations that demand the production of certain kinds of waste by producing a certain kind of goods (Cooper, 2010: 1119, quoting Gille, 2007; Evans, Campbell, and Murcott, 2013). Waste has given rise to vast economic sectors that provide income to multinational conglomerates, local authorities, and individual waste operators (O'Brien, 1999: 282). Seen from this perspective, waste is not a mere by-product of conspicuous consumption or the remainder of an excessive economy; rather, it exists in 'a society awash with rubbish: as a manufactured part of the world of goods involving labour, exchange, licensing, regulation and profiteering' (O'Brien, 1999: 286; Gille, 2010) and has become a fundamental and inalienable part of production.

However, as Gille makes clear, even though society entices people to acquire material goods, the 'consumer does not *make* garbage, nor do they make trash or have any choice in the materials they buy and turn into surplus stuff' (2010: 1050, italics in original). Following Lefebvre (2002 [1961]), people have been made part of, or have been implicated in, an economic mechanism in which they are supposed to create junk so that it can then be turned into a newly reusable resource. And even though attempts have been made to break through the principle of the built-in obsolescence of goods, this in itself does not guarantee a more efficient or ecologically less intrusive production process (Murray, Skene, and Haynes, 2017). Discarding possessions is often not the consumer's first choice; as Gregson, Metcalfe, and Crewe put it, 'to throw away (certain sorts of) things is an intrinsic part of contemporary being; a way of narrating ourselves through the presence *and* absence of consumer goods' (2007, 688; italics in original). Before deciding to get rid of possessions, people consider the stewardship of goods, or 'custodian behaviour' (Cherrier, 2010). They try to pass goods on to friends and/or relatives, hand them around to interested parties, or sell them, before contemplating the prospect of finally letting them enter the waste stream (Gregson, Metcalfe, and Crewe, 2007).

Waste in China

The confrontation with billowing amounts of solid waste is a fairly recent phenomenon in China. In pre-modern, Imperial days (until 1911-1912), superfluous goods were reappropriated and recycled endlessly until no more use could be found in or for them. In the Republican era (1912-1949) urban configurations expanded, and with them the urgent need to discard waste materials for which there was no further use, and which were produced by and for the fast-growing urban population. Scavenging, gleaning, and other occupational endeavours emerged as activities that were successful in both recycling matter and absorbing the surplus labour power that spilled into cities from the countryside. Even so, the total amount of waste increased, although it still seemed manageable (Dong, 2003). During the first three decades of the PRC (1949-1979), the nation was very much in the process of rebuilding from over 50 years of internal conflicts and wars. Recycling was hailed as a patriotic activity, or even a revolutionary duty. Politicization of the act of recycling was used to mobilize the people to contribute much-needed resources to the reconstruction of the nation. Indeed, in many respects the Chinese situation was similar to the post-World War II situation of Hungary described by Zsuzsa Gille, in which 'planners and workers alike hailed all garbage and by-products as "free" materials to be mobilized for the fulfilment of the plan [...] [T]he state implemented a vast infrastructure that registered, collected, redistributed, and ordered the reuse of both production and consumer wastes' (2010: 1056).

This changed with the Reform and Opening Policy formulated by Deng Xiaoping in the 1980s. In the relatively short time since then, China has discarded its planned economy and developed what it calls 'socialism with Chinese characteristics': producing for the global market while at the same time attempting to satisfy the desires, wants, and needs of the population (Russo, 2012). Alessandro Russo credits Deng's decision with setting the trend of global 'neoliberalism' (2012: 271); the late Arif Dirlik termed this new phenomenon in China 'post-socialism' (1989: 364), while David Harvey proposed the terms 'state-orchestrated capitalism' or 'neoliberalism with Chinese characteristics' (2003: 153; 2005: 120). Deng's policy liberated Chinese consumers from only having access to the products that were churned out according to the overarching production plan. Instead, they were urged to consume in order to support the development and growth of the economy. They were also able to consume more, because wages rose significantly. Producers no longer needed to simply fulfil the quota spelled out in the Five Year Plans; instead, they had to compete for a share of the market and, to do so, they had to seduce potential customers. By consuming more products, Chinese consumers demonstrated their patriotism. In other words, consuming was not merely about satisfying individual desires and wants, but also about serving the nation.

In many respects, Chinese society has followed the trajectory laid out in the theoretical discussions referred to in the preceding section. Industrial production turned into mass production of what consumers wanted, offering goods and services that were previously unavailable. The nation now faces solid waste disposal problems that are similar in many respects to those that other developing and developed nations in the world are grappling with. The urban infrastructure to deal with waste that was in place in the Maoist era has gradually been demolished, without alternatives being put in place or running. As various authors have established and made visible, China is now besieged by waste and threatens to be suffocated by it (Kao, 2011; Goldstein and others, 2011; Wang, 2011). It is an urgent problem: '[N]o other country has ever experienced the rapid growth in solid waste volume that China is facing now. The amount of refuse is growing annually by 8-10%, almost equal to the growth rate of GDP' (Zheng, Chen, and Craig, 2015: 67). The Chinese case is compelling because it is almost like a laboratory experiment that can be observed at a distance: at an incredible speed, the country is visibly experiencing a process that many other nations have also experienced or are currently experiencing, albeit at a slower speed and much more invisibly.

This study

This study focusses on the question of how China deals with waste. It is organized as a snapshot of the municipal solid waste (MSW) situation in one particular city in the People's Republic of China, i.e., Beijing, at the beginning of 2017. As the capital, Beijing is of course not China, and the city cannot be seen as representative of the nation as a whole. However, the policies and measures introduced in Beijing are monitored more closely than anywhere else in the country. As the capital, Beijing serves as a model for other cities to follow and often acts as a testing ground for new policies and approaches, with the result that the developments taking place there are analysed, described, and reported on more than developments in other Chinese urban areas. Thus, while keeping in mind that Beijing is or may be an exceptional case, the city forms a convenient research area. Moreover, the encroaching waste of Beijing has successfully placed the problems related to the final disposal of MSW visibly on the national agenda. Many other localities had been confronted by and trying to find solutions for MSW problems long before Beijing's waste siege became a topic of national interest, but they had failed to capture the attention of administrators, (environmental) non-governmental organizations (NGOs), or the domestic and foreign media.

The topics explored in the pages below do not focus on the technicalities of waste management processes. Rather, they are centred on the people who have to deal with waste that is not managed. China has expressed its intention to adopt the principles of the circular (closed loop) economy, yet the implementation of the legislative measures and policies involved in this decision encounters various problems of compliance at the lower levels of state organization. Per the adoption of the circular economy, the incineration of waste has been embraced as the ultimate solution for not only dealing with the waste itself, but also and more importantly producing and generating the amounts of energy (Waste to Energy, WtE) needed for continued economic development and growth while decreasing the burden on the environment. However, the process of burning the waste breaks the desired closed loop as potentially reusable resources are evaporated. Moreover, this 'incineration turn' is fiercely contested by some sections of Chinese society. The popular resistance to incineration seems unsuccessful and the attempts by already operating incinerators to neutralize fears of and suspicions about their operations fail. Successful incineration requires the careful sorting of waste. Newly emerging privately held recycling companies are increasingly taking part in the attempt to find solutions to this seemingly unsolvable problem. To make use of the preferential policy programmes formulated under the 'Internet Plus' plan launched in 2015, many of these companies have adopted an online-to-offline (O2O) strategy, supplementing their online presence with offline efforts. Has their emergence impacted urban residents? Indeed, how do residents dispose of their household garbage, and how and by whom is it collected? Some residents argue that O2O companies are merely new and more formalized manifestations of an informal waste collecting system that has existed almost without interruption through the ages. Migrant workers, who literally live on the margins of society, currently make up the main labour force of this informal system, and O2O companies see them as competitors. What do the waste pickers make of this turn of events? Various neighbourhood, municipal, and government institutions, as well as the O2O companies and environmental non-governmental organizations, have designed campaigns and programmes to inform and educate citizens and raise their awareness. The effectiveness of their efforts and those of the various environmental NGOs will also be scrutinized.

My work forms part of and was made financially possible through a broader project entitled Garbage Matters: a Comparative History of Waste in East Asia (financed by the Netherlands Organization for Scientific Research, project number 277-53-006), which sets out to examine waste as a social phenomenon in a number of contemporary East Asian nations (China,

Japan, South Korea, and Taiwan), and to explore the historical shifts behind the transformation of practices related to the 'production' and disposal of garbage since World War II.[1] In the East Asian countries under study, rising affluence, represented by growing levels of personal consumption, has played a critical role in this transformation. It has fuelled an endless expansion of the mass availability of consumer goods and has been accompanied by the overpowering encroachment by the waste generated by the food-processing and packaging industries. These and other developments have contributed to a metamorphosis of the daily practices of shopping and housework, as the culture of scarcity and the ethos of frugality have gradually given way to consumption-for-status and the veneration of material comfort and convenience (Strasser, 2003).

Data Collection and Constraints

A large part of the data that have been used in writing these chapters was collected during a period of intense fieldwork in Beijing during March-May 2017. Many of the impressions related to waste management and the informal sector of garbage pickers gained in that specific period expanded on earlier observations made during my very first visit to the city in 1980 and the many subsequent ones. I find it a treat to walk the streets and observe the daily behaviour of Beijingers; having a specific focus when observing them is even more rewarding. To be honest, however, my long-time interest in these topics cannot match the almost total immersion in the Chinese world of garbage of Joshua Goldstein, who has made it into the focus of his academic work (Goldstein, 2006; Hedrick, 2016), or Adam Minter's familiarity with the junk trade (2013a, 2013b, 2015). During my roaming through Beijing's neighbourhoods, I have been able to observe and compare the waste-related practices of the people of various districts of the city, ranging from those living in the sophisticated and intellectual Haidian and the sprawling and busy Chaoyang – the two districts where I have spent most of my time over the years – to the residents of Dongcheng, Xicheng, Shunyi, Changping, Fengtai, Tongzhou, and Shijingshan. When you hang out on street corners or near waste collection depots, it is almost natural to strike up a conversation with the other people hanging out there. Wherever I could, I engaged in conversations with as many people as possible, simply talking about their own attitudes and behaviours that touched upon waste and recycling.

1   https://www.garbagemattersproject.com/

Although such talk can hardly be called interviews, since they lack formality and a prepared questionnaire, they were very illuminating and often served as point of departure for more questions and, where necessary, alternative approaches to potential sources of information.

With the assistance and cooperation of the Department of Sociology of Peking University, in particular that of Professor and Associate Chair Liu Neng, I have been able to conduct a number of formal interviews, some 30 in all. My respondents were from different strata of society and included academics; residents of communities; representatives of non-governmental organizations (Green Beagle/Darwin Institute, Friends of Nature Beijing, Huan You Science and Technology, and Hong Chao); regular sanitation workers, waste collectors, and waste collection point bosses; and officials in the municipal administration at various levels. My plans to interview representatives of the O2O recycling start-ups unfortunately fell through. I have also had the opportunity to visit a number of waste collection points and the Gao'antun incinerator facility to witness their operations.

As Wu and Zhang have convincingly demonstrated in their ground-breaking study of 'garbage lives' in Beijing waste communities (2016), as well as many other studies (e.g., Van Rooij, 2012), the relatively short period of time I was able to allocate for research was not enough to win the trust of potential sources, particularly migrant workers. Chinese society is famously said to lack trust and to be implicitly suspicious of outsiders (Fukuyama, 1995; Tu et al., 2011). Shunned by civilized urban society and living an existence marked by precarity, migrant waste pickers tend to be even warier of any type of contact, keeping to themselves as much as possible. Winning their trust is one problem; being a foreigner and trying to win their trust enough that they will agree to become an informant is an even more complicated one, particularly when the foreigner does not have anything to offer in return (Ou, 2011). At a time when the Chinese government has whipped up an atmosphere of distrust towards foreigners by suggesting that they are all spies intent on defaming the country and/or stealing its secrets, being a foreigner made many potential sources suspicious and fall silent, all the more because they were already living on the margins (Brazil, 2018). Nonetheless, my attempts to shoot the breeze with ordinary Beijingers were relatively successful, although talking about trash turned out to be much more complicated.

The political climate as it has emerged since Xi Jinping's accession to power in 2012, with its suspicions and prosecution of corrupt behaviour and ideological non-conformity among officials and non-officials alike, has created a general fear of taking the initiative, of being seen as taking chances that others within the administrative and political hierarchies might find fault with. This was particularly noticeable when I attempted to reach out to the O2O companies that I had intended to be the centrepiece of this research project: Incom, Taoqibao, and Zaishenghuo. Despite formal letters of introduction from Peking University and other strategies to try and contact with these companies, not one was willing to be interviewed, stating that it was inconvenient for them. Among the reasons given for this unwillingness were the Two Meetings (Lianghui) sessions that convened in March 2017; the Belt and Road Forum (April 2017); and other events of an official and political nature. National events like Party meetings usually generate a tense atmosphere in the city, which is combined with increased surveillance and (visible and invisible) police presence. In spring 2017 Beijing was tenser than I had before experienced, with surprise ID-card inspections in the subways and other surveillance activities. The information collected through the interviews that I was able to conduct has been augmented with data and insights from an extensive and broad literature study on solid waste management, waste picking, urban development, precarity, etc., as well as topics more broadly related to Chinese society and politics.

Although Beijing's waste problems have been and continue to be re-searched by many others, until now most research has focused on the quantitative aspects of waste generation and the various methods of disposal that have been in place in various locations at various times (e.g., Dorn, Flamme and Nelles, 2010; Linzner and Salhofer, 2014). While a deep knowledge of these processes is essential, in my opinion serious engagement with the human factor of waste generation and disposal has largely been missing. What do ordinary people actually think about the waste they generate and the problems it creates for the city? How are their behavioural considerations influenced by rules and practices on a daily basis? Enquiries into the attitudes of those citizens who participate in demonstrations against incinerator plants and other acts of Not-In-My-Back-Yard (NIMBY) defiance (Johnson, 2013b; Wong, 2016; and others) have been published regularly in recent years, but this scholarly interest appears to be stoked more by the potentially subversive aspects of these civic actions than their underlying environmental causes or motivations. The voice of the ordinary person in the street continues to be rarely heard. In this study, I want to give this ordinary person a voice and hear where s/he stands. What are residents' attitudes towards their own polluting behaviour and what are their ideas about and expectations of waste management in the present and foreseeable future?

Structure of the study

In Chapter 1, I provide a sketch of how China, where 'everything was constantly recycled in a culture of thrift and poverty' (Dikötter, 2006: 14), particularly Beijing, evolved into its present state of overflowing waste. I look at the shrinking of state participation in dealing with waste and the emergence of an army of informal waste collectors that has been recruited or emerged from among the migrant workers who have moved to the cities. While the latter's contribution to waste management cannot be denied, their existence is hardly ever acknowledged in official sources, at least in a positive sense, and their ability to operate has been and continues to be made increasingly difficult. The chapter closes with an overview of the severity of the current waste situation, discussing the problem of collecting relevant and workable statics and arriving at an analysis of whether the situation is as serious as many sources suggest.

Chapter 2 discusses the desire to create a circular economy in China, as officially expressed by the government, noting that structural initiatives to bring this about are few and hard to find. Just as in many other countries, the Chinese government sees incineration as the most logical and most effective way forward (McKinsey & Company and Ocean Conservancy, 2015). On the one hand, incineration solves the problem of ever-growing amounts of waste in one fell sweep; on the other, burning garbage produces energy (WtE) that can be used for a variety of purposes. On a more symbolic level, the embrace of incineration technology shows the rest of the world how modern, developed, and evolved China is. Here, I also analyse why the state encounters so many problems in making people comply with its rules and regulations. Particularly relevant for the circular economy is the 'Internet Plus' initiative (State Council, 2015b) proposed by Premier Li Keqiang. Under 'Internet Plus', efforts were focused to 'integrate the mobile Internet, cloud computing, big data, and the Internet of Things with modern manufacturing, to encourage the healthy development of e-commerce, industrial networks, and Internet banking, and to guide Internet-based companies to increase their presence in the international market' (State Council, 2015b). 'Internet Plus' gave birth to the sharing economy (Lan et al., 2017), and allowed for the emergence of 'Internet Plus Recycle'. Under this heading, a number of garbage and recycling companies have made an appearance to fill the gap left by the disappearance of the recycling structure run by the government. Their defining characteristic is that they combine on-line and off-line (O2O) activities (Guo, 2016; Zhang, 2016), adding a contemporary twist to the centuries-old profession of recycling. Apart from offering to collect the garbage and recyclables from people's homes, many of these O2O-companies have branched out into other fields, such as offering small repairs and home refurbishing services (i.e., painting, electricity work). Where do O2O companies like Incom, Taoqibao, and Zaishenghuo fit in the stream of waste and its disposal? Do they recycle the waste they have collected themselves, or do they serve as middlemen? And why have they branched out to offer various other services, in effect becoming employers for part-time workers? Where do these companies acquire a workforce for jobs that are not related to garbage? Do they collaborate with urban employment bureaus or recruit on street corners? What does becoming an employee of such a company actually mean? Is it just about joining a work unit and wearing a uniform, or does it imply more, on a subconscious level? Does it give migrant workers an opportunity to become accepted members of the urban community? What influences, if any, are the activities of these companies having on the social fabric of Beijing? On the basis of my findings, I am forced to conclude that these various 'Internet Plus' initiatives that were hailed as the ultimate solution to the waste problem, have not realized their stated goals and may in fact have only served as a form of fashionable window-dressing through which to gain access to government subsidies and support.

Chapter 3 looks at Chinese urban residents, i.e., the people who produce garbage. I analyse the willingness of a number of people to classify and separate their garbage, and establish that awareness of the waste problem is not always matched with actual individual behaviour that seeks to contribute to a solution. Where do urban residents stand when it comes to dealing with the waste they produce? Are they aware that their ever-increasing consumption exacerbates the problem of waste disposal? Are they inclined to support the principles of the circular economy, and will they actually start to reduce, reuse, and recycle as much as they can? What are their ideas about the garbage retrieval services offered by the new O2O companies, and do they make use of them? It seems that the apparent failure of these net-based operators has further contributed to a broadly shared feeling of frustration among citizens concerning the government's handling of the waste situation. Many people want to actively contribute to improving the environment, but see no options by which they can put their energy to effective use. Are those willing to do so more in favour of solutions like waste incineration or are they becoming environmental warriors, mobilizing through civic action against secondary pollution and other potentially dangerous by-products? If they adopt a position that is more activist in nature, do they merely display NIMBY behaviour or are they seriously engaged in trying to find alternatives to improve the environment?

In Chapter 4, my focus is on the other side of the coin: the waste collectors. Before Beijing's waste is transported to the steadily decreasing number of landfills or increasing number of incinerator facilities on the outskirts of the city, it is painstakingly sorted for anything that might be valuable, which is taken out of the waste stream and sold. The people sorting through other people's garbage tend to be members of the vast army of migrant workers; in the process, they reduce the total amount of waste by 17-38 percent (Linzner and Salhofer, 2014: 905). Many of these migrant workers initially started working in the building trade, but once the scope and speed of construction work slowed down, they have become active as waste pickers, PET bottle collectors, or recyclers of other resources and raw materials. As Wu and Zhang (2016) have noted, this is not because they are unable to find alternative employment; rather, it is a conscious choice based on informality, independence, and a reasonable income. Due to developments beyond their control, the value of the recyclables they collect and sell is under pressure. With global oil prices dropping, it has become cheaper to produce new, fresh plastics and PET than recycle used materials for new resources. The Beijing municipal government continues to throw obstacles in their path, harassing them while they do their waste-picking work, pushing them further and further out of the city proper, and closing down their communities and workspaces. Moreover, the emergence of O2O companies may have changed the work conditions for such informal labourers. Because these companies display an organization, structure, and sense of sophistication that unorganized labourers lack, they appeal more to the administration. O2O companies try to offset the competition from the informal sector by co-opting it. But do the waste pickers consider employment by one of these companies and actually becoming an 'official' (or at least, a formally employed, officially recognized) garbage picker to be worth their while? Do they care whether such employment might be an option by which they could gain social capital (Prasad et al., 2012)? Is there competition between the pickers who have been co-opted by the O2O companies and the 'freelancers' who are still active, or between the garbage pickers from different companies who fight for access to the same recyclables? Does this new way of collecting recyclables have any influence on the often-witnessed, often-reported personal connections that have emerged over time between garbage-offering residents and garbage pickers (Prasad et al., 2012; Minter, 2013a, 2013b, 2015; Wu and Zhang, 2016)? In late 2017, the Beijing municipal government and others have started a number of concerted campaigns to drive out migrant workers and force them to return to the rural places from which they originate. What does this mean for the garbage collection, retrieval, and recycling sector? Or, to put it differently, who will take care of sorting the recyclables from the urban garbage if the informal collectors gradually return or are forced to return to the countryside?

Chapter 5 analyses the educational processes related to acquiring a recycling consciousness. People need to be informed of the problems involved in waste management. This includes not only being made aware of the need to produce less garbage, but also about less harmful ways to dispose of waste once it is generated. This calls for continuous educational efforts related to the need for garbage classification and separation, ways to reduce waste, etc. In the past ten to fifteen years, a socially acceptable and responsible attitude towards waste and garbage disposal has increasingly become widely associated with the concept of having *suzhi* ('human quality'). *Suzhi* is a marker with which one can define oneself or one's group in relation to others. One can learn or acquire *suzhi*, but it is essentially a quality that one is born with. With this concept of *suzhi*, urbanites set themselves apart from others, including the migrant workers who collect their waste. Yet the perception of having or lacking *suzhi* also plays a role when evaluating the behaviour of people from one's own social circle when it comes to garbage classification and separation. What educational projects on waste, garbage classification, and separation have been developed, where were they employed, and what results did they have? What means and media have been the most effective in bringing about changes in behaviours and attitudes (i.e., print media, public service announcements, special television programming)? Are small-scale local events more effective? How effective are regular educational postings on social media sites by the O2O recycling companies in changing behaviour? Who develops, designs, and produces these postings?

Chapter 6 looks at the work of NGOs and other voluntary environmental groups. Voluntary environmental organizations are very active when it comes to raising the consciousness of the population concerning garbage disposal, garbage separation, and the benefits these activities have for improving the living environment. Organizations such as Greenpeace International, Friends of Nature, Green Beagle/Darwin Institute, and many others are taking the lead in urban China in this respect. Yet non-governmental organizations (NGOs) in general, and even environmental NGOs (ENGOs), have to fight a battle on two fronts. In addition to their educational and activist work, they face stiff political and government obstruction to be able to be active in the first place. In the Xi Jinping era, the playing field for ENGOs has been reduced even further (Kostka and Zhang, 2018). The Chinese party-state generally considers organizational forms like NGOs to be a threat to its existence; in the best-case scenario, the government will confer on them a consultative role, or use them to reach certain parts of the population. Although this makes NGO work difficult in China, it is not impossible (Lu, 2007; Salmenkari, 2008; Wu and Chan, 2012). ENGOs are very active in a wide variety of environmental and garbage disposal-related initiatives and are able to draw on large numbers of volunteers from among school pupils, university students, and others. At the same time, ENGOs are very concerned about maintaining good relations with the government. As a result, they tend to shy away from actively supporting citizens' protests, whether they have a NIMBY character or otherwise. On the whole, the hands of Chinese ENGOs seem to be tied. While they have sprung into existence to give a voice to those among the population who are concerned about the way things are being done, make the people more aware, and improve environmental conditions, many residents remain under the impression that ENGOs are more interested in creating a united front with the government and improving the standing or status of the organization or its executives than in siding openly with citizens' demands, of whatever type. Yet it is only by adopting these strategies that ENGOs are able to circumvent the stringent regulations (Ho, 2007). In addition, green activists make avid use of informal networking with Party and state officials as well as with environmental scientists to raise the impact and effectiveness of their initiatives. What do NGOs concretely bring to the table, then? How can they deploy strategies that satisfy both the government and those that live in China's closely controlled civil society?

Chapter 7 discusses the politics of waste incineration. The people of Beijing seem to be generally positive about the possibilities of incinerator technology, particularly when that technology is imported from abroad, as is often but not always the case. Companies from Japan, Germany, France, Switzerland, and other countries have sold and transferred technology; Chinese incinerator designers include the Shenzhen-based Everbright Environment Co. Ltd and the New Century Energy and Environmental Protection Co. from Hangzhou (Chin, 2011). At the same time, there are deep-seated feelings of distrust towards the regulatory process that guides incineration, and towards the recycling factory managers and incinerator officials that are presently responsible for the plants. This distrust is largely based on incidents that have taken place in the recent past, leading to widely circulating and generally believed rumours about a lack of safety and the incompetence, corruption, and malfeasance of the officials. Some incinerator plants, including the Gao'antun Plant in Beijing, have developed a very open and above-board way of encountering the complaints, fears, concerns, and protests from the people living in their vicinity: they try to allay suspicions by opening their doors, making the processes taking place in their plants as visible as possible, creating the impression that they take the fears and complaints of people living in the neighbourhood seriously, organizing various neighbourhood activities, etc. Yet the suspicions and fears about the plants remain. Many citizens continue to see the activities organized by the incinerator facilities as simple attempts at whitewashing or hiding the true nature of what is happening inside the plants. Opposition to incineration plants and to the construction of other potentially polluting and environmentally hazardous factories such as PX (paraxylene) plants continues to exist, but actions and protests generally do not evolve beyond the NIMBY (Not-In-My-Back-Yard) level (Lee and Ho, 2014; Steinhardt and Wu, 2015; Zhu, 2017; Bondes and Johnson, 2017): people are mainly concerned about events taking place in their own neighbourhoods and do not really care about what happens in others, and there is no principled opposition or resistance against incineration as such (i.e., Not-In-Anybody's-Back-Yard). Is incineration the only way forward, or does it only solve part of the problem for Beijing? What is the state of the debate in Beijing and what is the level of public resistance? How has the construction of incineration facilities progressed, and has it contributed to the alleviation of the problem(s) caused by surplus garbage creation?

In the final chapter, I sum up my observations, provide recommendations, and identify venues and topics for future research.

Acknowledgements

This work would not have seen the light of day without the support and help of Ms. Xiao Boyu and Mr. Liu Yongbo, two promising graduates from the Sociology Department of PKU who went out of their way to act as my research assistants in 2015 and 2017, respectively. My 'laobaixing' friends of many years, which include Ms. Jiang Ping, Chen Yueming, Dai Jinquan, Xu Dashan, Tong Yue, Hao Peimin, Ms. Tang Ying, Ms. Shi Guizhi, Wu Erli, Li Yongqing, Ms. Yang Wei, Li Wei, Wu Yueli, and Zheng Shilin, opened the doors to the lives of 'ordinary' people for me, and allowed me to gain the perspective of the 'Beijinger', in so far as such a person actually exists.

Many friends and colleagues from academia went out of their way to help me gain access to information, sources, and informants, and/or to see things the correct way. They include Prof. Liu Neng (PKU), Prof. Ding Junjie (CUC), Prof. Liu Linqing (CUC), Prof. Wu Yongping (Tsinghua), Prof. Huang

Shengmin (CUC), Dr. Lin Hui (CUC), Dr. Hu Bo (CASS), Dr. Wang Yuqing (independent scholar), Ms. Mary Huang (PKU), Ou Ning (independent scholar), Dr. Yang Qingqing (MUC), Liu Chang (independent scholar), Prof. Wu Jing (PKU), Guan Shaoming (PKU), Liu Xinwei (Guangming ribao), Prof. Yuan Peng (CASS), Prof. Liu Jun (BFI), Prof. Zeng Guohua (CASS), Ms. Pang Yaran (PKU), Huang Tao (PKU), Liu Zhengnan, and many others.

Among other Beijing friends who have helped me out along the way over the years, Huang Hui, Dong Zhongchao, Ms. Xie Ying, Ms. Wu Min, Liu Yong, He Changhai, Cui Xinwei, Yang Shaoping, Guo Lei, Hu Dehua, Zhang Boyi, Cui Baodong, Luo Qilong, and Cheng Baogui surely must be mentioned.

The research for this publication took place under the auspices of the Garbage Matters: a Comparative History of Waste in East Asia Project, financed by the Netherlands Organization for Scientific Research, project number 277-53-006. I am grateful to the project leader, Prof. Katarzyna J. Cwiertka, for inviting me to take on the research of the China situation; and I am indebted to the other researchers involved in the project, Ms. Olivia Dung, Ms. Hyojin Pak, and Ms. Rebecca Tompkins, as well as Prof. Anne Murcott, for stimulating discussions and insights, and great dinner conversations.

1   Setting the scene – From Imperial to Present-Day Beijing

China has a long history of recycling; 'everything was constantly recycled in a culture of thrift and poverty' (Dikötter, 2006: 14). Frugality and thrift were part of the Confucian discourse that prevailed in a society of scarcity. This discourse advocated the need for a self-sufficient economy and considered wasteful habits to be part of a guilty lifestyle. Every material object could be turned into a commodity, and every commodity could be used and reused endlessly (Li S., 2002: 798). Goods that could no longer be recycled disintegrated on their own, since they were made from organic materials. Even human faeces was collected, mixed with food waste and other organic leftovers, and then dried and processed into fertilizer for use in the fields (Huang X., 2016). Recycling, sometimes by scavengers, took place both in the vast countryside and in the increasingly large, more urbanized areas that emerged along the Eastern seaboard in the late nineteenth and early twentieth centuries (Downs and Medina, 2000).

The Imperial City and the early Republic

Beijing was always a place where people consumed rather than produced, as opposed to other urban areas, such as Shanghai, that were more focused on commerce and industry. It was also a city where recycling traditionally made up a large part of the economic activities of its inhabitants. The city sprang up around the Imperial Palace and provided space for the vast administrative and bureaucratic structure that ruled over all of China. It had served as the capital since the last ethnic Chinese dynasty, the Ming (1368-1644), moved from Nanjing in 1421. These two institutions – the palace and the bureaucracy – and the people living in and working for them produced unfathomable amounts of waste on a daily basis, ranging from foodstuffs to clothing, from artefacts to paper, from broken crockery to night soil. While the city supplied space to the palace and bureaucracy, the latter's waste provided many if not most of the entrepreneurial classes of the city with much of their lifeblood (Goldstein, 2006). After the fall of the Empire in 1911-1912, recycling and reuse became even more pervasive in the lives of the inhabitants of Republican Beijing, becoming an essential aspect of its culture (Goldstein, 2006).

After the fall of the Empire and the founding of the Republic, the Imperial family and its vast retinue declined in numbers and importance, as did their power to consume. In the early years of the Republic, a political power vacuum emerged. The Republic's first President, Sun Yatsen, was ousted from office shortly after accepting the position. The commander-in-chief, Yuan Shikai, attempted to crown himself as the emperor of a new dynasty, but withdrew his claim after failing to gain the support of provincial military leaders and business interests. Without a strong and central leader, Beijing's governing authority was fiercely contested by military strongmen or warlords operating from local centres of power. Assuming that ruling the capital meant ruling over the nation, these military strongmen came and went, setting up new governments in Beijing, declaring themselves presidents of the Republic, and subsequently abandoning their offices – often after only a couple of months (Wu, 1991). Conquering the capital meant more than simply being crowned as the supreme leader; upon entering the city, each warlord gained access to foreign financial and material support and recognition as long as he held onto power, remained in Beijing, and was seen as representing the nation (Sheridan, 1975).

As a result of the reduced circumstances of the imperial elite and the arrivals and departures of temporary warlords and their armies, the stream of goods trickling down from high places to low, from wealthy to poor, grew and changed in character. Selling and bartering used goods and materials, including high-end articles like the often-priceless antiques and curios sold off by members of the Manchu nobility that had fallen on hard times, became a common economic activity. With the fall of the Empire, the gates were also opened for a flow of imported 'foreign' goods that satisfied the demands and expectations of the newly affluent cosmopolitan elite that emerged in China's cities around the same time (Zanasi, 2015). The indigenously produced goods that this urban elite rejected in favour of imports were picked up by others and refurbished. Together with other marketable goods that had been thrown out, collected, and reappropiated, these second-hand goods ended up at markets where they were sold to those who had less money (Dong, 2003; Dikötter, 2006; Goldstein, 2006). Many inhabitants of the city – those whose livelihoods had previously been inextricably bound to the types of employment that the system of imperial administration had required – now needed to find other ways to make a living. Waste picking became an option. In the stratified world of junk collecting, they were known as 'pole carriers', 'big basket toters', 'small drum beaters', or 'large drum beaters', each category of collectors on the look-out for different types of recyclables, weaving through the streets and alleys and calling people to sell or barter what they needed to get rid of (Goldstein, 2006: 266). Women, the elderly, and children were increasingly forced to turn to gleaning and picking garbage as a strategy for survival, as it provided them with food scraps and clothing. While Beijing and many of its people had fallen on hard times, studies of life in other late Imperial and early Republican cities and urban centres like Shanghai, Tianjin, and Chongqing show that conditions were not much better elsewhere (Gamble and Burgess, [c. 1921]; McIsaac, 2000; Rogaski, 2000; Henriot, 2013). In Tianjin, for example, the local Bureau Sanitaire rounded up the homeless, gave them uniforms, and had them clean the streets in return for food (Rogaski, 2000: 39). In these early, unsettled years of the Republic, the number of urban inhabitants kept growing, with people fleeing their homes in the war-torn and restive countryside and seeking shelter and subsistence in the cities. Others simply chose to try their luck in the bigger cities. Lacking the relevant training for urban occupations, most of these newcomers found work in the informal sector, where jobs ranged from pulling rickshaws to begging and waste picking (Strand, 1989; Lu, H., 1999; Goldstein, 2006). Although formally outsiders, these recent migrants were considered an integral part of the city's culture (Goldstein, 2006).

Nationalist Beiping

The dire straits in which the capital of the nation found itself became even worse at the end of the Northern Expedition (1925-1927), a military operation in which Sun's original Nationalist Party (Guomindang), now led by his successor Chiang Kaishek, joined forces with the relatively young and small Chinese Communist Party (CCP), which had been founded in 1921. The aim of this United Front was to defeat the warlords, some of whom had managed to conquer and rule over Beijing, and reunite China under centralized rule (Sheridan, 1975). In 1927, the United Front emerged victorious, having defeated most of its rivals on the battlefield or at the negotiating table. However, supported by foreign powers and conservative businesses and organized crime groups in China, the Nationalist Party turned on its erstwhile allies, the communists, in what became known as the Shanghai Massacre of 1927 (Sheridan, 1975). The Communist Party was forced underground and fled to the countryside, where it could regroup. With the nation once more under unified control, the Nationalist Party decided to move the capital of the Republic away from Beijing and establish it in Nanjing. This choice had grave consequences for Beijing and its inhabitants. No longer the political centre of China, no longer functioning as the beating heart of the bureaucratic apparatus that had managed the nation for centuries, the city was relegated to an inferior position in both a concrete and a symbolic sense. It could only fall back on its former splendour and try and market its spoils and remains (Dong, 2003). The city lost its original name in the process, to illustrate how much the new rulers apparently hated the old capital: it was renamed Beiping ('Northern Peace'), and ceased to be Beijing (the 'Northern Capital') (Strand, 1989). Under these circumstances, Beiping entrepreneurs and ordinary people had to come up with new strategies of survival, though they received ever-diminishing returns for their efforts. Recycling became the backbone of the wide variety of handicrafts and forms of household labour that emerged as a result (Ensmenger, Goldstein, and Mack, 2005).

Hostilities with Japan broke out in 1931, leading in 1937 to the full-blown anti-Japanese war that would become part of the Second World War. As Japanese military forces swept over China, Beijing/Beiping fell to the invaders in 1937. The city's now-inferior position was demonstrated by the lack of interest in its fate: most domestic and international attention focused on events in Shanghai (which the Japanese bombarded in 1937) and Nanjing (where the Nanjing massacre took place in December 1937, in which Japanese troops wantonly killed 300,000 civilians); on the withdrawal of the National-ist troops and government offices to safer destinations; and finally on the 1938 installation of the Nationalist government-in-exile in the inland city of Chongqing, Sichuan Province. However, the Japanese recognized the importance of the former capital, which they renamed Beijing and made the seat of the Provisional Government of the Republic of China, a puppet regime under Japanese control. This later merged with the puppet regime of the Wang Jingwei Government, which had previously been based in Nanjing (Wu H., 1991), to become the national government under Japanese occupation. Beijing's municipal archives are full of materials about the events that took place under the Japanese in the former capital, ranging from grand affairs to matters of daily life, that need more careful study and analysis than can be presented here. After Japan's capitulation in August 1945, Beijing's name once again reverted to Beiping. It became the centre of command for the Nationalist military actions during the civil war with the CCP, which broke out in the beginning of 1947. It also regained symbolic importance as the city where a growing number of popular protests took place against the corruption and inefficiency of Nationalist rule and against continued foreign presence and influence in China. With the Communist armies advancing victoriously and the Nationalist defences crumbling, Beiping was finally 'liberated' by the CCP in February 1949 without a single shot having been fired (Sheridan, 1975; Strauss, 2006).

People's Beijing

With the formal establishment of the People's Republic of China (PRC) on 1 October 1949, Beiping was again renamed Beijing. The newly installed government under the CCP immediately started to restore unified control over the country after the almost 20-year period of disruption caused by war and civil strife. With Beijing once more the glorious political capital of the nation as well as its symbolic centre, it was imperative that the city become the 'model of socialist urbanism for the rest of the nation, the embodiment of the criteria of socialist urban construction' (Lanza, 2018: 42). Beijing had to become a city of workers, with modern industries that had not before existed there, and a city where the needs of the masses could be satisfied. Amongst other consequences, this called for substantial numbers of new workers to be recruited from the countryside (Bergère, 2002: 106-109). To this end, urban space needed to be reconfigured and used in different ways; rural labour recruits needed to be resocialized into urban workers; the workers' daily lives and needs needed to be taken care of; and work, leisure, transportation, and residences needed to be connected and integrated on a scale that had not been known before (Bergère, 2002; Lanza, 2018: 43). Orderly urban management was high on the agenda of the national and municipal administration. Under Nationalist and Japanese rule, the departments responsible for running the city had been unable to turn around the results of decades of neglect and inaction: blocked sewers and drainpipes; impassable, muddy, and narrow roads; a public transportation system that functioned haphazardly; the inability to provide an uninterrupted provision of electricity; a lack of safe drinking water; and waste, organic and otherwise, everywhere (Wu H., 2005; Lanza, 2018: 44). The policies that set out to improve hygienic conditions included the formal recruitment of large numbers of recyclers who had been only informally active in the past, thereby consolidating their informal networks into more permanent, formally administered ones (Goldstein, 2006).

The initiatives undertaken to return urban society to a more ordered state included the restoration of sanitation services. From the early 1950s onward, two highly complex bureaucracies responsible for this process emerged in the city – one for the collection and disposal of household garbage; the other for buying and collecting recyclables and preparing them for processing

(Wang, Han, and Li, 2008). The latter bureaucracy also had to organize a system for the recovery of materials. The purpose of this system, as well as a number of top-down recycling campaigns, was to salvage as many resources as possible for use in national construction and industrialization (Li S., 2002). The recyclables were collected systematically, moving from individuals to neighbourhood redemption stations and networks, to larger district centres, and finally to regional recovery stations (Yang and Furedy, 1993). This organizational structure became a model that was copied on a national scale. The Beijing Municipal Scrap Recycling Company was formed in 1956 from the more than 7000 informal collectors, peddlers, and hawkers who had earlier plied their trades individually and were now reorganized into a formal work unit, a *danwei* (Ensmenger, Goldstein, and Mack, 2005; Goldstein, 2006). This process mirrored the policies of collectivization that were applied in other sectors. By joining the Company and becoming part of its work unit structure, scrap collectors actually attained an elevated position. They became state workers, with a national responsibility. This change in status provided these workers with the assurance of an urban household registration or *hukou*, which enabled them to enjoy the same subsidies and benefits that other Beijing residents had access to, including permanent employment, inexpensive housing, free medical care, and pensions (Gu, 2001: 92). In return for this administrative largesse, they had to hand over their savings, which served as the operating capital of the Company (Ensmenger, Goldstein, and Mack, 2005). The Company operated the collection points that dotted the neighbourhoods, as well as the stalls for buying goods that could not be reused anymore. Following the Beijing model, similar official redemption depots were set up in nearly every block in other cities. This nationwide organization of the urban waste management system worked very effectively (Li, S., 2002; Steuer et al., 2017). Collected industrial materials were transported directly to the factories for recycling, eliminating the intermediate levels of larger collection points and scrap markets. In this way, scrap collecting on this scale transformed into an important component of the industrial sector (Goldstein, 2006).

Campaigns

Administrative focus on the collecting and repurposing of scrap coincided with national drives to eliminate rats, fleas, and flies organized by the Ministry of Health. These campaign targets, including the creation of a garbage removal system, were all made part of the first Patriotic Hygiene Campaign in 1952, which directly linked cleanliness with health and modernity (Yang, N., 2004; Li, B., et al., 2015). Turning the people into healthy and modern citizens became explicit and important goals for the new nation. It enabled the state to demonstrate and project its strength and vigour. The Hygiene Campaign, as well as the many similar ones that unfolded in later years, had the additional function of disciplining the citizens, teaching them behaviours that were deemed to be in accordance with these new state ambitions. Part of the disciplining process covered the act of recycling: it was considered everyone's revolutionary, or even national, duty to contribute to collecting waste for recycling (Li S., 2002; Zhang and Wen, 2014; Steuer et al., 2017). The full-scale industrialization of China was one of the paramount objectives of the economic policies in the early 1950s, as evidenced by the First Five Year Plan (1953-1957; launched only in 1955) (Hooton, 1955; Muramatsu, 1955). The junk brought together at state waste redemption centres was used for the extraction of the raw materials and resources that were desperately needed by the resource-hungry industrial base (Steuer et al., 2017).

All Chinese cities routinely had recycling campaigns on the agenda, but in 1957 Beijing became the first city to announce that waste should be classified before its collection (Yang C., 2013: 176). The combination of waste recycling and classification campaigns called for the production of educational materials that would demonstrate and explain which types of materials could be recycled, as well as how and for what purpose. Most of the population undoubtedly had prior awareness and knowledge of classification and recycling, but the illustration of the purposes for which these materials could be reused – thereby showing the significant role they could play in the reconstruction of the national economy – was new (Li S., 2002; Chinese Posters Foundation, 2016). However, this did not mean that everybody complied. Even though a person could show his or her commitment to the revolution by handing over recyclables, for many urbanites in money-scarce Maoist China scrap represented wealth. People saw it as a nest egg for hard times, or used it as pocket money to go to the cinema or buy themselves a treat (Ensmenger, Goldstein, and Mack, 2005). By 1958, when the Great Leap Forward campaign was undertaken to propel China's economy to the same level as that of the United Kingdom, recycling turned into a positive act of citizenship that enabled everyone to participate even more in the great task of building the great new nation (MacFarquhar, 1987; Ensmenger, Goldstein, and Mack, 2005; Goldstein, 2006). By not actively and visibly taking part in this movement, individuals could run into trouble: such behaviour questioned one's motivation and commitment to the state's broader developmental goals.

The Great Leap Forward quickly failed to deliver on its promises of full-scale development, bounty, and abundance, and instead ended in a famine of catastrophic proportions. It was not only food that was lacking – though this resulted in the death of millions (Dikötter, 2010; Visser, 2016); the frenzied focus on the production of steel and other products for use in heavy industry resulted in the unavailability of almost every commodity. In the Great Leap period, the production of consumer goods came to an almost complete standstill in favour of the materials churned out by the primary industries. People were forced to use the things they had been able to acquire – or cling onto – in the past until the items could no longer be used. After the Great Leap, it took at least five years, until well into the 1960s, before food production was restored to a level at which starvation no longer threatened. Consumer goods continued to be scarce; they were often only available through the system of rationing that operated within the work units. Wei Jingsheng, a former Red Guard who turned into a human rights activist in the 1980s, provided an account of what this meant for the population well into the mid-1960s. In his autobiography, Wei recounts how he stumbled upon a group of beggars at a railway station during his 1966 train trip to the Northwest. This was at the time of the *chuanlian* or linking-up campaign, a movement that allowed Red Guards to freely travel the country to create links and exchange experiences with their fellow revolutionaries (Jian, Song, and Zhou, 2006; MacFarquhar and Schoenhals, 2006). Wei at first assumed that the beggars at the station were wearing dirty, dusty clothes. He recoiled when he discovered that they were local people who, as a result of the Great Leap Forward, not only lacked food but also basic clothing; they had to cover themselves with dust to hide their nudity (quoted in Fernandéz-Stembridge and Madsen, 2002: 207).

Recycling propaganda

Even the few possessions the people still owned after the Great Leap were coveted for recycling by the government. It can be seen as a sign of the times that the Beijing Municipal Scrap Recycling Company was renamed into the Beijing Municipal Resources Recycling Company in 1966. This renaming did justice to a reality in which junk was no longer scrap, but a resource (Ensmenger, Goldstein, and Mack, 2005; Goldstein, 2006). A number of propaganda posters and educational materials from the mid-1960s and early 1970s provide insights into the process of collecting and re-using goods. The posters lack exact publication details, but internal evidence allows them to

Illustration 1.1   Liaoning Old and Used Goods Collection Company (辽宁省废旧物资回收公司). **'Don't throw away broken leather shoes, recycle them to turn them into fertilizer'** ('破皮鞋莫扔掉回收加工造肥料')

Liaoning Old and Used Goods Collection Company, 1960s?, circulation unknown
IISH/Stefan R. Landsberger Collection. Photo © International Institute of Social History

be roughly dated. Only one of the posters shows a portrait of Mao Zedong on the wall or in the background; they provide no inspirational Mao Zedong quotes urging the people to comply; and none of the figures in the images wear buttons featuring Mao Zedong's likeness on their chests. All of these details point to the mid-1960s as the period of publication. Moreover, the artistic styles of the poster designs reflect the aesthetic preferences of the mid-1960s (Landsberger, 1995).

Two posters published by the Liaoning Old and Used Goods Collection Company demonstrate that dedicated organizations were at work to gather used goods in localities outside of the capital. The first poster focuses on old shoes, which could be turned into artificial fertilizer (Illustration 1.1). The developmental strategy at the time stressed the need to increase agricultural production to support industrialization, even if the image tells us that the fertilizer will be used to increase fruit production rather than increase growing staple foods. In the early 1960s, China was still recovering from the famine of 1958-1962. This poster's focus on handing in leather shoes is remarkable: most Chinese at the time would not have been able to afford leather shoes.

The second poster calls on the people to hand over their used glass containers, i.e., bottles, jars, 'cultural products', etc. (Illustration 1.2). The text suggests that the containers will not be melted down, but rather that each piece of glassware will be returned to the company that initially used it for packaging purposes, sterilized, and reused. Thus will the aim of thriftiness for the revolution be realized: fewer new glass containers will need to be produced and no raw materials will be wasted. Ten items of recycled or reused glassware make it possible to save five pounds of coal and more than a pound of sodium carbonate.

The third poster was published by the Shandong Provincial Local Products Company (Illustration 1.3). It illustrates in detail the wide variety of consumer goods that could be handed in and the products they would be turned into after reprocessing. Many of the examples project a sense of turning the old into something new and modern: old rubber boots and tires reappear as trainers, for example. Such posters were used to make people realize that many goods they might consider useless and not worth handing over, such as human hair, could actually support the goal of development. The lower left-hand corner of the poster shows what a collection point may have looked like at the time, though this example is located in the countryside. The chart on the wall shows the buying prices of different goods, pointing to the existence of a well-functioning collection and distribution system that connected supply and demand on a national scale. In the present age,

Illustration 1.2   Liaoning Old and Used Goods Collection Company (辽宁省废旧
物资回收公司)**. 'Recycle old bottles to use them again, increase
production and be thrifty for the revolution'** ('旧瓶回收能利用增产
节约为革命')

Liaoning Old and Used Goods Collection Company, 1960s?, circulation unknown
IISH/Stefan R. Landsberger Collection. Photo © International Institute of Social History

Illustration 1.3   Shandong Provincial Local Products Company (山东省土产公司).
**'Eagerly sell old and useless materials to support the construction
of socialism'** ('踊跃出售废旧物资支援社会主义建设')

Shandong Provincial Local Products Company, 1960s?, circulation unknown
IISH/Stefan R. Landsberger Collection. Photo © International Institute of Social History

such price lists are still used at collection points, but they now quote global prices for recyclables – an indication of the transnational character that recycling now has assumed (Medina, 2011; Hunwick, 2015; Wu and Zhang, 2016). A final interesting detail about this print is the name of its producer, the Shandong Provincial Local Products Company. This name provides no indication that it specialized in recycling or collecting junk.

The fourth poster was published by the Beijing Municipal Office for Saving Paper (Illustration 1.4). It urges the population to economize their use of paper, and to collect and re-use old paper boxes, to further support the revolution. This poster shows which types of paper the Office was interested in. It explains, among other things, that one ton of recycled paper boxes would save the nation 500 kilos of coal and 350 Watts of electricity. It also lists the addresses and telephone numbers of the factories and offices that would buy used paper and details which places specialized in which specific types of used paper. Had this poster been published later than the mid-1960s, the final product the recycled boxes would have been used to produce paper to produce more editions of the *Selected Quotations of Mao Zedong*.

Illustration 1.4   Beijing Municipal Office for Saving Paper (北京市节约板纸办公室). **'Economize on paper for the revolution, collect and re-use old paper boxes'** ('为革命节约板纸回收复用旧纸箱')

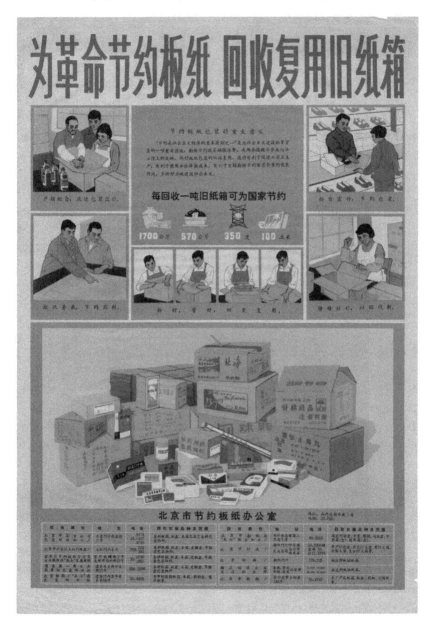

Beijing Municipal Office for Saving Paper, date of publication unknown, circulation unknown
IISH/Stefan R. Landsberger Collection. Photo © International Institute of Social History

When, by the mid-1960s, the spectre of mass starvation had been over-turned, the CCP devised another political movement. This was to make sure that the Chinese still followed the Party's commands despite the failures of the Great Leap. This new movement, the Great Proletarian Cultural Revolution (1966-1976), would again throw Chinese society into turmoil. No one was safe from potential prosecution, neither the high and mighty like Vice-Presidents (for example Liu Shaoqi) and Army Marshalls (such as Peng Dehuai), nor the people on the bottom rungs of the social structure (MacFarquhar and Schoenhals, 2006). One victim from the last group was Shi Chuangxian (1915-1975), a Beijing night soil collector who had been recognized and honoured as a national model worker since 1949. Shi was able to parlay this symbolic distinction into membership in the CCP and a political and administrative career as a representative of the National People's Congress; he also was received by many State leaders, including Mao Zedong and Liu Shaoqi, and praised for his work. A 1959 photograph that had appeared in the official CCP newspaper *People's Daily*, in which Shi was shaking hands with Liu Shaoqi, became his undoing. Once Liu had been driven out of office by the Red Guards in 1967, Shi was persecuted for his alleged connections with Liu and beaten so seriously that he became paralyzed. Despite efforts to provide him with care, he died in 1975. Only after Mao's death was Shi officially rehabilitated (Jian, Song, and Zhou, 2006: 256-257).

In the first two years of the Cultural Revolution, the Red Guards, students, and high school pupils who had been mobilized to support the ideas of Mao Zedong ransacked the homes of many individuals who were suspected of having bourgeois or otherwise politically incorrect lifestyles. Many personal belongings of the victims of these raiding parties, ranging from works of art, calligraphy, and antiques to clothes, furniture, books, and other housewares, ended up in the streets, ready for collecting. Although there certainly may have been waste collectors who were willing to pick these up, there was also political risk attached to trying to sell these discarded goods. The same risks pertained to the potential customers, who would shrink from buying goods that were considered tainted by a counter-revolutionary provenance.

Despite the seemingly chaotic political situation at the time, however, the enterprise of recycling and salvaging as much as possible continued as before. One last poster, published in the early 1970s by the Beijing Municipal Scrap Recycling Company for *neibu* ('internal' company) use, calls on people to hand over as much scrap metal and other waste materials as possible (Illustration 1.5). The triumvirate of a worker, peasant, and soldier clearly dates it to the Cultural Revolution period; so does the slogan on top – a Mao

Illustration 1.5   Beijing Municipal Scrap Recycling Company (北京市物资回收公司). **'Strive to collect scrap metal and other waste materials!'** ('大力回收废钢铁及其他废旧物资!')

Beijing Municipal Scrap Recycling Company, early 1970s, circulation unknown (internal use)
IISH/Stefan R. Landsberger Collection. Photo © International Institute of Social History

Zedong quote that reads 'Prepare for struggle, prepare for famine, for the people' – and the copies of the *Selected Works of Mao Zedong* that the worker and peasant hold up. Like the other posters discussed in this chapter, this poster explains how the junk that is handed in saves virgin resources and energy in the production of ploughs, minivans, and bicycles. On the whole, this image has a more militaristic and aggressive tone than the other posters.

Beijing under Reform

Once the Cultural Revolution ended formally in 1976 and the (economic) Reform and Opening Policies of Mao's successor Deng Xiaoping started to gain traction in the 1980s, the lives of most Chinese changed dramatically. These changes started in the countryside, where rural communes were decollectivized and individual farmers were given their own plots of land to cultivate, in return for turning over specified amounts of the harvested crops for predetermined prices. What the individual farmer produced over the quota could be sold on the newly organized farmers' markets, and the proceeds could be pocketed. This in effect meant that harder work resulted in higher incomes, a philosophy that industry also came to embrace. While the rural areas were the first to profit from this structural change in development strategies, the effects gradually spread to the urban areas and gathered in speed and scope. One of the hallmarks of the era, was that people no longer were dependent on their work units for the distribution of foodstuffs and other commodities but could individually buy them as long as they could pay for them. In response, factories started churning out consumer goods in dizzying abundance. The bulk of industrial production shifted away from heavy industry and related projects to the wants and needs of the people. In a relatively short period of time, urban wages rose many times over and consumption increased to unprecedented levels, trickling down from the elites to the rest of the urban population. As a result, China quickly changed into a consumer society with characteristics that are quite similar to Western ones, and has come to face problems associated with consumerism that many other developing and developed nations in the world are also grappling with. This makes China an interesting object of study: many of the problems that have emerged in developed countries in the past can be seen emerging there, almost in front of our eyes.

The generation of municipal solid waste (MSW) is a prime example of such a problem. In a relatively short period of time, waste generation increased at an unprecedented speed and the question of how to dispose of the waste did not meet any quick and satisfactory answers. Where in the three decades from 1950 to 1980 MSW production in Beijing grew from 1500 to 3000 tons per day, it grew to 5800 tons per day in 1985 and 9050 tons per day in 1990. The make-up of the MSW changed in the process too. In 1950-1980, it consisted largely of coal ash, mixed with food or kitchen waste; in later decades, cans, plastics, paper, etc., came to make up the major part of MSW. The disappearance of coal ash from Beijing's garbage was the result of the prohibition of the use of coal for heating and cooking purposes in the inner city; over time, the definition of the boundaries of the inner city have changed and now include the area within the Fourth Ring Road (Goldstein, 2006; McKinsey & Company and Ocean Conservancy, 2015). While the general make-up of Chinese garbage is quite similar to that of other developed and developing countries – keeping in mind that the percentages of the components vary – it is also considered extraordinarily 'wet' because of its high food waste content (Goldstein, 2016; Minter, 2014; Interview with Hong Chao, 2017).

The time-honoured practice of the general population recycling (or selling) waste materials continued under the Reform policies. However, it became increasingly difficult for people to find buyers interested in their junk. The employees of the Beijing Municipal Resources Recycling Company had declined in numbers and as a result were faced with more work than before. But the definition of the Company's work also had changed: it now considered sorting through trash to be too degrading. Instead, the Company maintained its focus and the bulk of its activities on the recycling of industrial wastes, such as iron, aluminium, and copper, over which it held a monopoly. This allowed the Company to reserve this lucrative trade for itself and for industry insiders with a proper Beijing registration, thereby excluding migrant labour (Goldstein, 2006). When real estate became one of the growth poles in urban China under market reform, the Beijing Municipal Resources Recycling Company withdrew further from its original activities. Instead, the Company turned into an entrepreneur, forming alliances with local governments, local developers, and (foreign and domestic) investors to engage in real estate development activities that could produce more profit than dealing in scrap. More profits also positively influenced the relative standing and clout of the Company and its managers and employees. As a consequence, the vast and finely organized network of waste collecting points in Beijing, which had come to number some 2000 by the 1980s, rapidly disintegrated and almost completely disappeared, leaving only six in operation by 1998 (Ensmenger, Goldstein, and Mack, 2005; Goldstein, 2016). Although landownership rights continued to be vested in the government (or the state), the Company could sell the land use or land-lease rights to the plots that had been allocated for the recycling stations and put these spaces to more profitable use by initiating urban construction projects.

This market of land use or land-lease rights started after 1988 and really took off in Beijing in 1992 (Goldman, 2003; Kostka, 2014). The rights to the recycling stations were sold off to real estate developers, who turned them into the more aesthetically pleasing and better paying high-rise apartment buildings that were erected all over the city with astonishing speed. This brought the developers significant profits as well, since hardly any residents of these spaces needed to be resettled (Kostka, 2014). Urban development activities were further stimulated and covered by the Old and Dilapidated Housing Redevelopment Programme started in Beijing in the 1990s, under which more than one-quarter of inner-city dwellings were deemed to require urgent attention due to structural instability and severe deterioration (Shin H., 2009). Many of the recycling stations could be found in these parts of the city. Rather than renovating them, they were torn down and replaced, often with the financing of foreign investors (Shin H., 2009: 2821). This practice extended beyond the inner city and came to embrace all of Beijing's districts, displacing many of the original inhabitants and forcing them ever further away, onto cheap suburban land at the margins of the city. This extraordinary building boom that started in the 1990s has transformed the city beyond recognition (Broudehoux, 2007).

Enter the waste pickers

During this same period, the amounts of garbage and recyclable waste, the daily life waste produced by Beijing's citizens, mushroomed. This was the result of the steadily expanding size of the population, the growth of its disposable income, and its patterns of increased consumption as well as greater opportunities to consume in the form of new types of businesses such as restaurants, shops, shopping malls, and hotels. A form of conspicuous consumption once more took root in urban China as it had in the 1920s and 1930s, but this time consumption became more and more status-driven and -defining.

In the vacuum left by the withdrawal of the Beijing Municipal Resources Recycling Company, a new phenomenon emerged. This was made up of informal waste pickers, most of them migrants from the countryside who moved to the cities, including Beijing, to try to find a way to make a living. They had become redundant in their places of origin; overemployment meant that there were no job opportunities in agriculture. The migration processes and employment histories of these informal workers are discussed in more detail in Chapter 4. These waste pickers now concentrate on collecting, separating, and selling the recyclable parts of household waste, taking over the work the Company is no longer interested in. They have connected with the new emerging businesses, something that the Company was not able to do. They transport all the junk through a hierarchical network, from residential communities to collection depots to markets, recycling factories, and even state-owned enterprises. As the waste is transported upward, it attains ever higher levels of purity and value. And they do all this without any legal basis. According to Goldstein, quoting Beijing waste administration sources, this network handled more than 1 million tons of materials, generating more than 1 billion yuan in profits, in 2001 (Goldstein, 2006: 281).

China's waste – attempting to assess amounts

According to statistics published by the former Ministry of Environmental Protection, now the Ministry of Ecology and Environment (MEE), the 246 largest cities in China produced 186 million tons of household waste in 2015. Of this waste, 90 percent was buried in landfills or incinerated without being properly sorted. This raises questions about the effectiveness of the sorting activities of the informal sector, and runs counter to the generally held belief that most recyclables are taken out of the waste stream by informal sorters (Interview with Huan You, 2017). The same statistics indicate that Beijing city produced 7.9 million tons of garbage in the same year, of which 44 percent was burned (China Youth News, 2016). By 2020, Beijing city plans to burn 67 percent of its household waste (Yuan and Li, 2017). By 2025, the World Bank estimates that China's urban population will have grown to 1.4 billion people, who will generate 500 million tons of MSW per year (Hoornweg and Bhada-Tata, 2012: 80). The China National Renewable Energy Centre estimates that by 2020, 277 million tons of collected MSW will need to be disposed of, reaching 369 million tons by 2030 and 409 million tons by 2050 (Energy Research Institute of Academy of Macroeconomic Research and National Development and Reform Commission, 2017: 316).

Yet these figures only form a part of a much larger and much opaquer picture. Over time, Chinese statistics have increasingly been treated with suspicions about their reliability, in particular when it comes to economic data (Wallace, 2014). The figures regarding waste generation and treatment are fragmentary and notoriously incomplete (Linzner and Salhofer, 2014: 897). There are a number of reasons for this. Some authors insist that Chinese statistics in general are based on 'guesstimates' and that the figures quoted above attempt to paint a rosy picture about official garbage disposal activities within a region (Liu, Zhang, and Bi, 2012; Wallace, 2014; Lo, 2015). Such a favourable depiction can have a positive effect on the evaluation of the performance of local administrations or individual officials and can influence promotions or demotions, rewards and punishments. The information-gathering structure is outdated and fragmented, sometimes dating back to the system of economic planning, and leaves out many relevant variables. In other instances, the officials responsible for reporting the data are the same ones who collected them in the first place (Liu, Zhang, and Bi, 2012; Wallace, 2014; Lo, 2015). In addition, most of the often-quoted figures on waste and waste treatment are only based on the situation in some 660 cities, excluding the entire rural population of China. While this population is declining from its previous majority of the total population to less than 50 percent as a result of the ongoing process of urbanization, the fact that a large proportion of the population is not included in such calculations raises questions (Dorn, Flamme, and Nelles, 2012; Albores, Petridis, and Dey, 2016: 266). Moreover, the data do not report the amounts of waste recorded at the point of generation and omit the quantities of recyclable waste that have already been taken out of MSW by the informal sorters. According to several researchers, some of the recycling companies stick to overreporting the amount of waste they recycle, as it allows them to apply for more government subsidies (Chen, Geng, and Fujita, 2009: 38; Yang C., 2013: 177). To get a firmer grip on the problem, and in line with the dictum 'you can't manage what you can't measure', a method of more systematically gathering quantitative data and information is needed to monitor and control the situation (Naustdalslid, 2014: 309, 310). As Naustdalslid argued in his analysis of the plans to implement the circular economy, '[S]ince most of the practical efforts to promote CE are in the form of pilot programs and demonstration projects, evaluation criteria and indicators for measuring success and failure become particularly important' (2014: 309). The same applies to the various waste-related efforts that are planned or currently taking place.

The last obstacle encountered when attempting to estimate China's waste is the question of whether the waste one sees is actually produced by China itself. In a number of publications, Adam Minter (2013a: 85-88, 2013b, 2015) has shown how the countries that imported China's export products, in particular the United States, have sold and exported their waste and garbage back to China. China had an enormous appetite for resources, and many developed countries who consume finished Chinese products had more recyclable waste than they care to recycle. Junk could be shipped to China at relatively low cost, using the containers that would otherwise return empty to be loaded with consumer goods again. For the shipping companies, this would have meant an unacceptable loss. As a result, a perfect loop of exported products and imported junk came into existence. According to government figures, China imported 49.6 million metric tons of waste in 2015 (Agence France Press, 2018). In the summer of 2017, however, the CCP Central Committee and State Council adopted the '2018-2020 Action Plan for Full Implementation of the "Implementation Plan for Banning Foreign Wastes and Advancing the Institutional Reforms on the Management over Importation of Solid Wastes"', making it clear that it was no longer interested in recycling other nations' garbage. The ban on the import of foreign waste, which took effect on 1 January 2018, is to be fully implemented by 2020 (State Council, 2017; MEE, 2018). This decision was not only based on China's estimate that it produced enough garbage of its own, or that it no longer needed to recycle waste to have access to resources; it was also seen as an opportunity to further stimulate the development of a circular economy. The ban also added opportunities to further regulate the domestic recycling sector, phase out informal waste pickers, and make the waste disposal system healthier. These conditions further dovetailed with the decision to expand the system of completely incinerating garbage that the government committed to in the 13th Five Year Plan (2015-2020).

2   The circular economy in China

In the Chinese view, the concept of the circular economy is based on the three principles of Reduce, Reuse, and Recycle. The rapid growth and industrialization of China over the past 40 years or so has given rise to more and more detrimental environmental pollution, which compounds concerns about increasing waste and decreasing resources. The circular economy is meant to serve as a response to these environmental challenges and to reconcile economic and ecological imperatives by decoupling economic growth from natural resource depletion and environmental degradation (Qi et al., 2016; Murray, Skene, and Haynes, 2017). In their comparative research on the definitions of 'circular economy' that circulate in China and Europe, McDowall et al. (2017) have established that the Chinese plans focus more on the combatting of pollution, sustainable development, and ecological civilization, while Europe is more concerned about solutions for waste and the opportunities for industry these solutions may bring (Blomsma and Brennan, 2017). By striving to rebalance economic development while also accounting for social and environmental objectives, China is attempting to achieve a state of harmonious development (Naustdalslid, 2014). The combination of ecological civilization – defined as China's long-term vision of sustainable development – and the circular economy fits the political narrative of 2002, when both these concepts were first introduced (Geall, 2015a). More concretely, elements such as resource efficiency, green technologies, and waste management were paired with calls to build a resource-saving and environmentally friendly industry and society (McDowall et al., 2017). The policy of creating a harmonious society – including a harmonious approach to development – is very much associated with the Hu Jintao era (2002-2012), but Hu's successor Xi Jinping has also continued to support it (Naustdalslid, 2014; Marinelli, 2018). The policy of creating an ecological civilization was incorporated into the CCP Charter at the 18th National Party Congress in 2012 and has been considered a key element of China's national development strategy since then (Kuhn, 2016; Marinelli, 2018).

The idea of the circular economy was first brought up by scholars in 1998 (Yuan, Bi, and Moriguichi, 2006). The circular economy as a strategy was formally adopted as a policy starting in 2002, but it took time to formulate the concrete measures by which it would operate. Environmental regulatory regimes based on concepts of cleaner production, industrial ecology, and ecological modernization that were already in place in Germany (i.e., The Waste Avoidance and Management Act, 2002), Japan (i.e., The Basic Law for

Establishing a Sound Material-cycle Society, 2002), and elsewhere, inspired the Chinese government to set up a number of pilot projects to explore how to implement the circular economy strategy (Someno, 2014; Blomsma and Brennan, 2017; Murray, Skene, and Haynes, 2017). After these tests, the Circular Economy Promotion Law (also known as the Recycling Economy Promotion Law) was adopted in 2009. It aimed to improve resource utilization efficiency, protect and improve the environment, and realize sustainable development (National People's Congress Standing Committee, 2008: art. 1; Liu, Zhang, and Bi, 2012; Naustdalslid, 2014). Environmentalists saw the law as concrete proof that all the talk about the necessity of establishing an ecological civilization through a circular economy was now being matched with deeds. In the years following the adoption of this general law, specific legislation was added to clarify where the rules were to be enforced, including the sectors of municipal waste, industrial waste, and wastewater (Qi et al., 2016: 41-44). While the plans and strategies were committed to the concepts of reducing, reusing, and recycling, they put less emphasis on measures that would influence patterns of consumption. Instead, specific manufacturing sectors and potential policies were scrutinized to increase efficiency and reduce waste and pollution in manufacturing (McDowall et al., 2017). To facilitate conversion to a circular economy, a Special Fund for Circular Economic Development was set up in 2012, but who can apply and which projects are funded remains unclear (Nelles et al., 2017). China's adoption of the circular economy has been a boost for the adoption of elements of the concept elsewhere (Murray, Skene, and Haynes, 2017: 369). Yet, the implementation of the circular economy is less developed than its theory and principles may suggest, and the process of transforming the economy to be more sustainable has not, in fact, made much headway (Yuan, Bi, and Moriguichi 2006; Naustdalslid, 2014: 308).

Incineration as part of the circular economy

Accompanying the adoption of the circular economy, incineration has been embraced as the ultimate solution for the waste problem. Indeed, incineration is said to effectively achieve more than a 90 percent reduction of waste volume (Y. Li et al., 2015: 234). However, as Martin Melosi has argued, incineration should not be seen as a potential disposal panacea: it is just one of several disposal options that are available, and what is needed is a determination of under what circumstances incineration serves which disposal needs best (Melosi, 1996: 41, 40). Incinerators not only deal with the waste itself, but more importantly, they can also produce and generate the amounts of energy (Waste to Energy, WtE) that are needed for continued economic development and growth while decreasing the burden on the environment. This makes incineration more attractive than other waste-disposal methods. The amount of power generated by incineration since the beginning of the 12th Five-Year-Plan has steadily increased: the amount for Shanghai rose from 24 to 58 percent, for Beijing from 13 to 45 percent, and for Guangdong from 34 to 56 percent (Y. Li et al., 2015: 239). By generating electricity during waste disposal, the air pollution caused by (coal burning) power plants can be reduced considerably. This is an attractive alternative, as the blue skies push that has been underway for some time has resulted in the closing down of many of the smaller and older coal burning power plants still in operation around and in urban areas (Kennedy and Chen, 2018). China is one of, if not the largest global energy consumers and relies to a large extent on the importation of energy resources from abroad to satisfy demand (Albores, Petridis, and Dey, 2016). Since the supply of energy resources from abroad is influenced by political stability elsewhere, as well as global pricing policies, the domestic production of energy resources creates a certain level of self-sufficiency. The Chinese domestic energy demand has shown a steady increase over the past decades, and there are no indications that it is slowing down.

In the process of burning the waste, however, the desired closed loop of the circular economy is broken because any potentially reusable resources left in the waste are evaporated. This calls for a pre-sorting stage prior to incineration, which however negates the idea that all waste can simply be chucked into the incinerator and be done with. This pre-sorting or recycling stage is where 'Internet Plus Recycle' companies with O2O-business models, discussed in more detail below, develop their activities. Better sorting and recycling practices of the general residents would also improve incineration effectiveness (Wan, Chen, and Craig, 2015).

The number of incineration plants nationwide has nearly doubled, from 238 in 2010 to 514 in 2015 (Yuan and Li, 2017), but due to the fast rate of expansion a credible and up-to-date inventory of all of the MSW incinera-tor plants is lacking (J.-W. Lu et al., 2017). Given the urgency of the waste problem, the disposal of garbage through incineration is seen as one of the pillar industries in the current 13th Five Year Plan (2015-2020); as a result, tens of billions of yuan are being poured into the construction of an infrastructure of incinerator plants all over the country (Compilation and Translation Bureau, 2016; Goldstein, 2016). This should double China's incineration capacity by 2020 (Yuan and Li, 2017). The central government uses the promise of state funding for incineration plants as an incentive to prod local governments into improving their environmental performance (Heberer and Senz, 2011). Densely populated large cities with scarce land resources see WtE incinerators as the most desirable waste processing technology because of its high efficiency, minimal land requirement, and significant impact on the reduction of solid mass. Incineration technology is seen by some as the only alternative to landfills (Wan, Chen, and Craig, 2015).

The demand for incinerator facilities has created a frenzied buyers' market for companies developing the technology and equipment. As early as 2009, research indicated that over one-half of the global orders for new waste incineration factories came from China (Balkan, 2012). The companies try to outbid each other with exceptional offers and claims of performance. Allegedly, bribes are also offered to the officials who are responsible for acquisition. These rumoured acts have come under threat as a result of the anti-corruption campaign that started in 2014-2015, which has brought down many corrupt officials at all levels. To undercut the competition, Chinese companies are said to offer unbelievably low burn rates. This suggests that the required air-pollution control systems, such as flue-gas filtering, are not installed, or that other cost-cutting methods have been employed (Balkan, 2012; J.-W. Lu et al., 2017). In 2018, the newly reorganized Ministry of Ecology and Environment once more addressed this pressing issue of flaunting emission standards, with the promulgation of the 'Action Plan for Straightening out the Municipal Solid Waste Incineration Power Generators to Meet Emission Standards' (MEE, 2018).

Implementation of laws and regulations and their constraints

In April 2015, the Party Central Committee promulgated Central Document Number 12: 'Opinions of the Central Committee of the Communist Party of China and the State Council on Further Promoting the Development of Ecological Civilization' (Central Committee, 2015). The document signalled a break with the past in that it announced a shift in policy priorities from economic growth to sustainable development. Although classified as mere 'Opinions', this document did not voice high-level intentions but instead proposed actual standards, mechanisms, and assessments to improve policy implementation (Kuhn, 2016; Geall, 2015a, 2015b, 2015c). These 'Opinions' defined the circular economy in terms of reusing and recycling resources, as opposed to simply extracting them, using them for manufacturing, and then consigning them to waste. The policy document further illustrated the government's commitment to the policy and assured that the 'Opinions' would have a major influence on the formulation of the 13th Five Year Plan for the period 2015-2020 (Geall, 2015b). The 'Opinions' reflect Xi Jinping's ambition to reroute the course of economic development towards a more environmentally friendly direction. This is illustrated by the inclusion of the lines 'the great value of lucid waters and lush mountains'. Xi first uttered these in 2005 while serving as a CCP chief in Zhejiang Province (Geall, 2015a; Zhejiang Provincial Committee 2016). Xi's remarks, now known as the 'Two Mountains Theory' of 'clear waters and lush mountains are invaluable assets that are comparable to gold and silver', have since become an important principle for discussing Chinese development (Ministry of Foreign Affairs, 2016). The document signalled to the rest of the administrative and political structures that the highest levels of both Party and State administration were serious about bringing ecological improvements and that the lower authorities needed to shape up (Liu et al., 2015).

Ran Ran (2013) has argued that China's environmental governance shows a paradox. While there is great awareness of environmental problems at the central levels of government and a comprehensive and modern set of environmental legislation has been promulgated to pursue sustainable development and environmental progress, none of this seems to have produced noticeable structural effects in practice. This state of affairs stems from the assumption held by the Chinese authorities that once they have at their disposal the correct factual knowledge about the state of affairs in society, adequate measures and policies will follow and implementation will be assured (Naustdalslid, 2014: 309, 310). To put it more strongly, in the words of a Chinese legal scholar, China's green laws are useless, as they are symbolic (as quoted in Simões, 2016).

The compliance and implementation of government decisions are obstructed by a number of interrelated causes, with China's extremely decentralized governing structure the root problem. Central government initiatives and laws tend to be 'showcase politics': ambiguous, inexplicit, unclear, vague, and overly reliant on good intentions rather than on hard measures, goals, and timetables that can be measured. Such policy expressions are intended to focus the attention of the bureaucracy on a particular policy area, and the mobilization of the commitments and focus of the bureaucracy itself is the most important target, rather than the implementation of the policies (Strauss, 2006: 896; Kennedy and Chen, 2018). This is particularly visible when it comes to environmental rules and regulations (Ran, 2013).

The relative lack of success of the implementation of environmental legislation is further illustrated by the ineffectiveness, caused by its low status and weak bureaucratic influence, of the institution that was responsible for environmental affairs. The former State Environmental Protection Bureau, which was responsible for all of the environment, used to be a mere bureau under the authority of the Ministry of Urban and Rural Construction. In 1988, it became a vice-ministerial level department and in 1998 it was upgraded to become the State Environmental Protection Administration, situated directly under the State Council but not a component of it. As a result of its low bureaucratic level, it was unable to press its case in confrontations with stronger, economically more important ministries. It was only after its elevation to full ministerial status in 2007 that the Ministry of Environmental Protection (MEP) gained importance and bureaucratic clout (Heberer and Senz, 2011; Qi et al., 2016). In 2018, the MEP was reorganized and renamed the Ministry of Ecology and Environment (MEE), which concentrated the various lower-level bureaucracies active in the fields of ecology and environment into one centralized structure (MEE, 2018). As concerns about the environment have increased, so has the power of the organization, in the process taking over responsibilities that were previously overseen by the National Development and Reform Commission (Kostka and Zhang, 2018).

The fact that China is a diverse country with widely different local characteristics and levels of prosperity has created a tendency for the central government to formulate standards rather than give explicit policy directives. In other words, the government more or less sets the agenda and leaves it to the local levels to bargain and modify the central policies into concrete measures (Heberer and Senz, 2011). As a result of decentralization, the Centre has retained only weak control over policy implementation and, as many scholars have pointed out (Ran, 2013; Heberer and Senz, 2011; Kostka, 2014; Kostka and Mol, 2013; Lo, 2015), there are few incentives for local officials to actually take sides when conflicts of interest arise between national regulations and local stakeholders. Moreover, the central authorities are prone to adopt symbolic legislation and displays of exemplary behaviour without first assessing whether their policies can be implemented. One example is the Circular Economy Promotion Law mentioned above, a policy document that sets out principles but at the same time lacks concrete standards and mechanisms. The officials who are responsible for its implementation do not understand its procedures and aims, and there is no broad popular support for the principle of the circular economy because the general public is hardly aware of it (Naustdalslid, 2014: 310, 311).

The Chinese ratification of the Paris Agreement on 3 September 2016, part of the United Nations Framework Convention on Climate Change, which was adopted on 12 December 2015, is another example of this symbolic posturing

(Kuhn, 2016).[2] On the basis of its past opposition to global environmental initiatives and as one of the major contributors to climate change, the Chinese government's support for the Agreement came as a surprise. The country was globally praised for its clear-sightedness, gaining considerable symbolic capital in the process. Many, mainly foreign, observers had to reconsider their often-outspoken critical position towards China when it comes to the environment. The praise only grew in intensity once the United States announced its withdrawal from the Agreement in 2017 and China was expected to take over as the global climate change leader (Kostka and Zhang, 2018). However, the real burden of implementing the conditions of the Agreement was shifted onto the shoulders of the lower and local bureaucratic levels, which have to come up with ways to make the changes that will enable China to comply with the goals it has underwritten.

Performance Evaluation

Usually, Chinese government officials at all levels occupy their positions for a period of five years, although a 3.5-4 year period is also frequently encountered (Kostka, 2014). It is every bureaucrat's aim to be promoted to a higher, or better, position. Occupational improvement can take the form of higher rank; additional wages or bonus payments; a better working environment (i.e., promotion to a bigger city, a bigger department, etc.); more access to allowances like subsidized housing; and so on (Kostka, 2014). To be eligible for promotion, every official has to submit to the so-called Cadres Performance Evaluation System, which is run by the Organization Department of the Party. This system specifies sets of targets that local cadres must fulfil. In very general terms, the targets include morality, capability, diligence, performance, and probity, but most are very concrete (Ran, 2013). The Organization Department can add targets and mark them as either 'hard' or 'soft'. Hard targets include GDP growth, maintaining social stability, successfully upholding the One Child Policy (until 2016), etc. Most of these hard targets additionally have 'veto power': if one of them is not accomplished, all other targets, even when successfully met, are annulled (Heberer and Senz, 2011; Kostka, 2014). Economic development was and still is seen as most important, while environmental measures were often seen as problematizing or impeding growth; environmental targets were 'soft' and tended to have fairly low priority in the evaluation rankings. A

2   http://unfccc.int/paris_agreement/items/9485.php

complicating factor in the enforcement of environmental measures is that measurements of the outcomes and effects of some hard target policies, like GDP growth, are easier to establish than those of soft target policies. It is relatively easy to prove that no demonstrations or riots have occurred in one's jurisdiction, or that the birth-rate has declined. The results of a greening policy, on the other hand, may take years to become visible. The push in recent years to increase the number of 'blue days', days when air pollution is successfully reduced, in major cities, results in numbers that cannot be disputed (Kennedy and Chen, 2018). However, the effects of policies that need a longer time to come to fruition will not be recorded on the score sheet of the officials who have enforced them, and their successors will reap the benefits instead. Understandably, officials are wary of working to improve the reputation of others and this generally impedes the implementation of environmental policies (Kostka, 2014; Eaton and Kostka, 2014).

Moreover, the decentralized nature of the political system means that the institutions that are tasked with implementing policies suffer from overlapping or contradictory responsibilities while also facing strong boundaries between administrative entities. Local Environmental Protection Bureaus (EPBs), for example, are responsible for supervising and enforcing the environmental regulations handed down from the higher levels, but they are dependent on other sections of the local government for their financial resources and staff (Liu, Zhang, and Bi, 2012; Heberer and Senz, 2011). It is clear that EPBs can be forced to temporize or simply refrain from enforcing policies when other sections of the administrative structure consider them detrimental to realizing their own goals. Development and Reform Commissions at the local levels are formally responsible for master planning for ecological improvement and environmental protection, but their top priorities are effectively industry development and investment in public infrastructure developments. Land Resource Bureaus are formally responsible for land, mineral, and marine resource protection, but they see collecting land and mineral resource fees as their main priority (Ran, 2013; Kostka, 2014). These bureaucratic impediments make it extremely difficult to enforce environmental policies that run counter to the development agenda that drives the locality. However, since the adoption of the 11th Five Year Plan in 2006, a number of environmental targets have been upgraded to 'hard' status level. This should ensure that these targets are implemented (see Lo, 2015: 154, 155 for the specific targets). Added to this is the growing conviction, also mentioned in Central Document Number 12 ('Opinions'), that an official's non-compliance with state-set environmental targets should be permanently included in his/her personnel file, thus influencing his/her future career chances (Geall, 2015a, 2015c; Zhang, 2015; Kuhn, 2016). Note that the 'Opinions' merely express a conviction; no concrete steps have yet been taken.

Despite these changes, complying with central government policy orders that are identified as hard targets is often at loggerheads with local demands, situations, and peculiarities. There are many occasions where it is more prudent to accommodate these local needs rather than implement central policy. In the worst cases, policies are only implemented selectively, or formally, without any concrete actions undertaken (Heberer and Senz, 2011; Kostka, 2014; Lo, 2015; Interview with Hong Chao, 2017).

Compliance with laws and regulations and its constraints

The decentralized administrative system currently in place in China lacks the means of enforcing the strict implementation of environmental policies and laws and has no tools for deterring lower levels of government from breaking the rules. How can compliance be enforced? This question is not only relevant for the behaviour of different layers of bureaucracy and the individuals manning them: it also has implications, as we shall see in later chapters, for all of society, down to the level of individual residents of urban residential communities. Individual behaviours regarding garbage classification and separation need to change to bring about the central government's goal of creating an ecological civilization. To put the question differently: if and when there are rules and regulations, why are they not implemented and internalized by officials and civilians alike?

As Benjamin van Rooij and others have shown, the study of compliance in China has only recently been taken up. Although the topic has been researched exhaustively in Western contexts, it is too early to tell whether the results of these studies can also apply elsewhere (Yan, van Rooij, and van der Heijden, 2016; Van Rooij et al., 2017). The extensive research on compliance in various instances of ecological legislation and related topics that Van Rooij and his collaborators have undertaken in recent years has unearthed some interesting findings that are pertinent to the question of individual compliance. The reasons for bureaucratic inertia are not difficult to find. More interesting is the lack of compliance with rules, regulations, programmes, and projects on the part of urban residents – the people who should be on the receiving end of policies, which apparently have not been handed down well from the centre to the locality. The preceding discussion has illustrated how a lack of enforcement or deterrence for non-compliance has allowed the responsible administrative layers to disregard promptings from higher levels in the bureaucracy; in practice, the same process applies to the residents. Despite intensive propaganda and other activities designed to influence and change their attitudes, residents seem unwilling and/or unable to internalize behaviour that is beneficial for improving the environment.

The previously mentioned compliance researchers Yan, van Rooij, and van der Heijden have established that, apart from forcing compliance through enforcement and deterrence measures, the factor of voluntary compliance is crucial for determining behaviours. Voluntary compliance is influenced by three conditions. The first concerns the operational costs and benefits of compliance and violation beyond punishment. None of the environmental plans and pilot projects organized until now mention a threat of punishment or provide opportunities to shame shirkers at the lowest levels for not acting in accordance with legislation. The second condition refers to the legitimacy of compliant behaviour. Legitimacy can be broken into four constituent factors: social norms, personal morals, perceived sense of duty, and procedural justice. Social norms are what educational campaigns attempt to instil: when others are engaging in this sort of behaviour, I will comply, as it is probably right. Personal morals touch upon whether one personally considers an action to be morally right. A perceived sense of duty comes from the feeling that one has to act in a certain way because it is seen as a general duty. The last factor, procedural justice, is related to whether one perceives the suggested behaviour as fair. The final condition is the capacity to obey the law. Again, present policies lack any pressure to obey any law on the level of the individual (Yan, van Rooij, and van der Heijden, 2016: 211-212).

The decision of an administrative level to not comply with rules and regulations handed down from higher levels is not influenced by aspects of voluntary compliance. The organizational unit is constrained on all sides to behave and follow the other units. When it comes to the individualized behaviour of urbanites, it is clear that much more is needed to bring about voluntary compliance than merely publishing laws and regulations or putting up posters, no matter which administrative level is doing so. Implementation of a third-party surveillance mechanism would be a first step. Recently, a system of rewards and punishments has been announced through the Social Credit System that will be implemented in coming years (Chen, Geng, and Fujita, 2009; State Council, 2014). However, the difficulty to make individuals comply also points to the urgent need for more and different types of education and persuasion, almost to the level of the individual urban resident. Most citizens currently have no idea what is expected of them. The ongoing efforts to educate and persuade residents to start classifying and recycling their waste are taken up more extensively in Chapter 5.

Internet Plus

Premier Li Keqiang proposed the 'Internet Plus' initiative in 2015 (State Council, 2015b). This was intended to add impetus to the creation of the circular economy and at the same time create a new engine for economic growth. The Internet Plus plan was adopted to 'integrate the mobile Internet, cloud computing, big data, and the Internet of Things with modern manufacturing, to encourage the healthy development of e-commerce, industrial networks, and Internet banking, and to guide Internet-based companies to increase their presence in the international market' (State Council, 2015b). On the one hand, the initiative was meant to facilitate the emergence of new, Silicon Valley-type high-tech industries and entrepreneurs. Another aim was to help older industries turn to the Internet to find new relevance and possibly stop them from folding altogether. The direction set out in the Internet Plus plan was further supported by one of the decisions at the 2016 Central Economic Work Conference, which stated that it was 'imperative to adopt and implement the new development concept, namely the concept of innovative, coordinated, green, open and shared development' (Wu, 2016).

Li Keqiang's Internet Plus initiative signalled to both the administrative levels and commercial sectors, particularly producers of information and communications technology (ICT), that the telecommunications network should be made more comprehensive and faster nationwide, allowing more people and more companies to become active online. Specifically, the relevant ministries and administrations were tasked with further establishing an internet infrastructure and formulating a 'Broadband China Plan' (Hou, 2015; Davidson, 2015; Fung Business Intelligence Centre, 2016). The better, cheaper, and faster telecommunications coverage that was promised benefitted the indigenous smartphone producers. Formerly known as *shanzhai* ('copy-cat') producers, they were now stimulated to come up with models that performed better than the much more expensive foreign brands, thus persuading more people to use them (Yang, 2016). By increasing the use of smartphones, more specially tailored applications could be developed to help stimulate another development strategy, the Made in China 2025 plan (State Council, 2015c). This plan envisions how China in 2025 will become technologically self-sufficient and turn into a manufacturing superpower by transitioning away from labour-intensive industries and toward advanced industries like robotics, advanced information technology, aviation, and new energy vehicles (Wübbeke et al., 2016; Laskai, 2018).

Many smaller entrepreneurs have come up with innovative initiatives under the wings of these programmes. Some of the outcomes of the strategy that are regularly praised by the Chinese media are initiatives to boost the rural economy. For example, the setting up of e-commerce production villages in the countryside is regularly featured in media reports and mentioned positively as a success. These villages have created employment and marketing opportunities, producing goods and selling products through online retail platforms like the one developed by the technology and retail giant Alibaba (Davidson, 2015). Looking at the situation in 2017 Beijing, the so-called O2O phenomenon (referring to online-to-offline activities, 线上 线下) has given birth to a wide range of business models and approaches to solve all sorts of social and consumer problems. Many if not most acquisitions in shops, most restaurant cheques, and most supermarket bills can now be paid through one's smartphone, by using one of the many online payment systems (wallet apps) that have been developed by companies such as Alibaba (i.e., AliPay). Ordering a taxi through a smartphone app, comparable to the way the Uber Company is operating in Western countries, was very much the thing to do around 2015 but seems to have fallen out of favour again.

The sharing economy

The most visible manifestation of Internet Plus has been the growth of the sharing economy, which gives substance to the reduction of consumption and waste generation as envisioned in circular economy philosophy. Numerous companies (OfO, MoBike, BlueGogo, etc.) started to operate dockless bicycle-sharing services in 2016, putting brightly coloured bicycles in the street that can be unlocked for use by scanning a QR-code attached to their seats or frames and making payment through one's smartphone (Lan et al., 2017).[3] This has proven a huge success: in 2017, MoBike was awarded the first China Social Enterprise Award for its innovative business model and approach to solving social problems; the Awards are organized by the Social Investment Forum and China Social Enterprise (Bhandari, 2017; Carlson, 2017). The easy availability of bikes has brought about a true cycling craze in many urban areas. Renting a bike has put a curb on the use of automobiles; it has offered a solution for urban congestion as well as air pollution; and has

3   QR-code, or Quick Response code, is a machine-readable optical label.

even contributed to the fight against urban China's expanding waistlines by making people exercise more. By tracking each unlocked bike for the duration of its use through Internet of Things (IoT) technology, huge amounts of data are generated every second that can be used for urban planning purposes, such as establishing where gaps exist in public transit (Yin and Tan, 2017). Bike renting has come to account for a third major form of public transport (Yin and Tan, 2017; Fieldnotes 2017).

A negative side effect of this sharing craze is that the huge numbers of yellow (OfO), orange (MoBike), blue (BlueGogo), and other-hued bikes that are left behind on streets and street corners have created a major problem for urban management. They have made negotiating pedestrian routes and sidewalks practically impossible, particularly for those who are physically or visually impaired (Bhandari, 2017; Yin and Tan, 2017; Fieldnotes, 2017). Many urbanites criticize others for carelessly disposing of the bikes, grumbling about their users' lack of quality, i.e., just leaving their bikes wherever they please rather than parking them correctly in designated areas; moreover, they are not pleased about the waste created by wrecked and otherwise broken bicycles that are left behind to rust (Interviews, 2017; Taylor, 2018). The infatuation with the sharing economy has not stopped at bicycles or Uber-like taxis. Adventurous entrepreneurs have succeeded in persuading venture capitalists to finance plans for companies that offer umbrellas to share, or even shared smartphone battery packs (Chen, 2017; Lu H., 2017). As a result of this massive turn towards smartphone use for calling services, particularly in metropolitan areas, people's lives have become intimately, almost literally, linked to their phones; their eyes are glued to their tiny screens, not looking up. This has given rise to the moniker *ditou zu* ('bent-headed tribe'). While some argue that this phenomenon is largely concentrated among the so-called post-1980s and -1990s generation of youngsters (*baling hou, jiuling hou*, the Chinese designation for millennials), one does not need to look very hard to see that older people have also become heavy smartphone users (Fieldnotes, 2015, 2017; Interviews, 2017).

While the Internet Plus plan clearly serves as a national agenda for development, it also leads to ever more moments and opportunities for the consumption of services and goods. The mushrooming services sector that runs on a 24/7 schedule has become one of the new domains where surplus labour is absorbed, usually in the form of delivery persons or couriers; large quantities of migrants supply this surplus labour. It is also provided by the new urban underclasses that have been emerging, which consist of laid-off employees of former state-owned enterprises, or graduates from vocational schools or even universities who have been unable to find more fitting employment (Engebretsen, 2013; Evans, 2014; Liu, 2016). One of the greatest ironies of developing China may well be that while the population across the board has been able to go to schools and universities in larger numbers than ever before (Murphy, 2004), has become better educated than ever before, the supply of jobs for these educated and/or trained newcomers has not increased at the same speed.

It is hard not to notice the army of delivery persons who have become an indispensable part of the service sector. They are usually men, very occasionally women, and predominantly young, although one also encounters elderly delivery persons. They are often dressed in eye-catching work outfits and many make use of O2O bicycles to make better speed on the congested roads to deliver their orders. Food-delivery companies have been teaming up with O2O bike companies. Street corners are occupied by tricycle vanlets, operated by delivery men and women who pile their parcels on the street, waiting for customers who have ordered goods online, from their homes, their workplaces, or other places, to come pick them up. Car owners have their cars washed and waxed by mobile car-wash-and-wax teams, which use tricycle vanlets or small delivery vans to carry the equipment and liquids needed for the cleaning (Norcliffe, 2011). The number of couriers delivering take-out orders to the hungry defies counting. With China's ever-expanding economy, the possibilities and opportunities for the service industry seem endless, but in the end, it all comes down to more consumption; to spending more money; to individuals rendering services for others higher up on the economic ladder; and to producing more garbage and waste that needs to be dealt with. Parcels are wrapped in extravagant amounts of wrapping and packing tape; all food deliveries come packed in plastic or styrofoam.

Internet Plus Recycle

In the newly created niche of 'Internet Plus Ecology', institutions and companies have been invited to enhance the dynamic monitoring of resources and the environment, develop intelligent environmental protection, develop and complete the system of collecting and reusing waste and used resources, and build an online waste-trading system (State Council, 2015b). Using the term 'Internet Plus Recycle', a number of garbage disposal and recycling companies have become active in recent years that combine online and offline activities, hence their inclusion as O2O-companies. They attempt to fill the gap left by the disappearance of the formal recycling structure, collecting recyclables and transporting them further upstream and competing with the informal waste collectors. Some of these companies are involved in the actual recycling process itself (Goldstein, 2006; Guo, 2016; Zhang, 2016). With the help of Chinese venture capitalist investors, many of these companies have developed smartphone applications to reach the citizens of Beijing and other urban areas (NDRC, 2016). Through these apps, they create dedicated user networks for their services, which include the collection of recyclables at the residents' doorsteps. Among these companies are Beijing Incom Resources Recovery Company (active in Beijing); GEM (Tianjin, Shenzhen, Wuhan); Zaishenghuo (Beijing); Beijing Zailai Keji (Beijing); and Baidu Recycle 2:0 (Beijing/Tianjin). As the last initiative focuses exclusively on the recycling of e-waste and smartphones, I have excluded it from this analysis.

The apps these companies have developed all look different, but they share some characteristics. They link waste producers with waste collectors or waste collecting systems. These apps show the user what his/her garbage is worth at any time: plastic and PET bottles are calculated per item, paper and cardboard by weight. The system of set prices means that haggling is no longer necessary. This may be beneficial for the person offering the recyclables, but not necessarily for the junk collector. The value of large objects like TV sets or washing machines usually has to be negotiated with the person picking them up, but not all companies adhere to that policy and have set prices. After indicating that scrap can be picked up, a waste collector, sometimes formally employed by the company behind the app, contacts the person offering the scrap; makes an appointment for the pick-up; and transfers the amount of money due through the options offered by mobile technology in the form of a wallet app (Zheng J., 2017). In the whole procedure, no cash money changes hands, further blocking opportunities for bargaining and embezzlement. Some companies that operate in a similar fashion do not transfer money at all, but have set up a system for saving points for consumer goods. On the basis of the number of points collected, one can order products that are then delivered by the same people who retrieve the recyclables. Some have argued that this merger of scavenging and mobile technologies results in the Uberization[4] of an occupation that for decades has been the lifeline for millions of people moving from the countryside into urban areas (Pasquier, 2015). In the section of the smartphone application that contains the personal details of the user (telephone number, address, etc.), the total amounts of goods that have been recycled are stored. This

4   Uberization: 'Changing the market for a service by introducing a different way of buying or using it, especially using mobile technology', *Cambridge Dictionary* (https://dictionary. cambridge.org/dictionary/english/uberize), accessed 3 January 2017.

big data can be used in Internet of Things-schemes, for example providing information to municipal sanitation departments for the improvement of garbage collection routines – but it can also be integrated at some point into the Social Credit System that the government wants to implement.

GEM, a company based in Wuhan with branches in Tianjin and Shenzhen, was one of the pioneers in the development of smartphone applications. In June 2014, GEM launched an app called Huishouge (Recycle Brother) in the presence of Yu Zhengsheng, member of the Chinese Communist Party Political Bureau and Chairman of the Chinese People's Political Consultative Conference (Guo, 2016).[5] The support of such high-ranking officials serves as an indication of how in line these types of operations are with government policies. To perfect the operation of its app, GEM focused on the construction and improvement of online platforms, the layout of offline recycling systems, the construction of logistics and warehousing, the construction of R&D teams and data centres, the promotion of online and offline advertising, and the exploration of environmental sanitation. Beyond that, the company has also invested in expanding its recycling network by building or integrating existing waste stations, setting up urban transit stations, and configuring logistics transportation between front-end recycling and transfer stations (Huishouge, 2017b). Since it runs services in Guangzhou, Shenzhen, and Wuhan GEM is involved in activities that are geographically more extensive and therefore covers more potential customers/users than the other companies that are concentrated in the metropolitan areas of Beijing or Shanghai, and yet it is not as well-known as the others (Zhang, 2016). In July 2016, GEM was recognized by the National Development and Reform Commission as a 'Top 100 Internet Plus' initiative on the basis of its unique business model and excellent recycling service (Huishouge, 2016).

Many other recycling companies have opted for a similar online-to-offline strategy. In 2016, an estimated 100 of them were registered (Zhang, 2016). In Shanghai, AlaHB[6] started in 2010 as a community-based recycling platform supported by a state-owned recycling company (Tong and Tao, 2016). Green Earth,[7] a community-based garbage sorting company in Chengdu, Sichuan Province, set up an information system for residential communities. With the slogan 'Trash is Cash', Green Earth provides incentives to registered members according to their performance in garbage sorting by tracking their discarding behaviour with RFID codes on the garbage bags they hand

5   http://www.huishouge.cn/
6   http://www.alahb.com/IndexNew.aspx
7   http://www.lvsediqiu.com/

in (Tong and Tao, 2016).[8] In Beijing, one of the best-known O2O companies is the Beijing Incom (Yingchuang) Resources Recovery Recycling Co. Ltd., which has developed a smartphone app named Bangdaojia. Its high domestic and foreign visibility is the result of sophisticated publicity policies. Taoqibao (developed by Beijing Zailai Keji Youxian gongsi) and Zai Shenghuo (developed by Anewliving), amongst others, are competing with Incom. At the time of writing (Spring 2018), the websites of these latter two companies were no longer active. Despite the rise and fall of some companies, the business model remains favoured. The Green Cat (Lümao) app was launched in Dongcheng District in April 2017 with the support of the Beijing Municipal Commission of City Management, which reposted the news on its website (Liu, 2017). Green Cat initially plans to service 20,000 households in the district, but over time and depending on the success of negotiations with the various departments involved, aims to expand that number tenfold.

All of these companies operate apps that are comparable to GEM's Huishouge, but some of them have looked for other ways to tap into the potential of this online-to-offline market and expand their operations (Illustration 2.1). Zai Shenghuo in particular has ventured into new fields that may serve as an example for others – and in fact some of them have already followed suit. Styling itself 'China's leading community life service platform', Zai Shenghuo has moved beyond collecting scrap and recycling activities. The company, originally established in Beijing in 2014 and expanding its operations to Shanghai in 2015, has added cleaning, odd jobs (extractor hood cleaning, knife sharpening, smartphone repairs, installing WiFi, etc.), and other maintenance services to its list of available activities. It also has a section for laundry services and offers the services of housekeepers, wet nurses, nannies, and nurses to care for the elderly. The first of these expanded services were announced with considerable fanfare around the 1 October festivities in 2016, and they now form such a large part of the company's activities that the original recycling work occupies a secondary position.

With garbage recycling becoming a billions-of-yuan activity, competition between these companies is fierce and becoming sharper all the time. Simultaneously, not all of the companies presenting slick apps and even slicker websites can stand close scrutiny. During Spring 2017, rumours made the rounds that the Taoqibao company had turned unresponsive. The app no longer worked and the general-service 400 phone number the company used for communicating with customers was not answered. The reason(s) Taoqibao

8   Radio-frequency identification (RFID) uses electromagnetic fields to automatically identify and track tags attached to objects.

Illustration 2.1   Screenshots by the author of the Bangdaojia (left, Incom) and Zaishenghuo (right, Anewliving) apps showing some of their services. May 2018

seemed to have disappeared remained unclear. More and more frequently, information circulated that Taoqibao's board of directors had vanished, taking the investors' money with them; according to people familiar with Beijing business practices, this has happened often (Interviews, 2017). A year later, in 2018, the Taoqibao website has ceased operations and been taken over by a company called Men's Heaven (男人天堂); the site is unresponsive.

All the companies mentioned here combine internet-related functions with activities in the physical world. Yet many questions regarding their operations remain. How successful are they, or have they been? Do the services they offer address consumer needs? How effectively have they harnessed the Internet? For some, waste collecting is no longer their main focus, but do those that continue specialize in specific types of garbage? And where does the waste go to, i.e., is the garbage recycled, incinerated, or landfilled, or all of the above? Have they merely become middlemen in a new, modern guise? Do the companies have an environmental consciousness and do they cooperate with environmental organizations, or are only looking for a profit (or dividends for their investors)?

Case study of Beijing Incom Resources Recovery Recycling

One of the first Chinese companies that made use of big data in relation to recycling and the environment was the Beijing Incom (Yingchuang) Resources Recovery Recycling Co. Ltd. Set up in 2008, it boasts of offering a 'Total Solution for Intelligent Solid Waste Recovery Machine and Recycling System' (Beijing Incom Resources Recovery Recycling Co, N.d.). As part of the INCOM Resources Recovery Co., Ltd., which is in essence a regenerated bottle-grade PET chip manufacturer, the company's aim is to set up an overall recycling system with a stress on intelligent solid waste recovery and recycling. As far as obtaining the raw materials for recovery, Incom has developed reverse vending machines (RVMs) for the Chinese market that have until present focused only on retrieving plastic beverage bottles. On the outside, these RVMs look similar to the machines in many supermarkets in Western nations for returning glass or PET bottles that carry a deposit. One important difference with Western practices is the refund procedure. After all, PET bottles in general are produced for singular use and are intended to be thrown away, not returned. At the same time, in Chinese popular thinking, all trash has value, even when it can no longer be used (Minter, 2013a, 2013b, 2015). To entice people to leave their valuable trash in the machines operated by Incom, an incentive is offered: disposing of a plastic bottle will earn the recycler credit on his or her public transport card, or extra mobile phone minutes. The Incom RVM is able to 'buy back' PET bottles in three sizes (0.5 litre and smaller sizes; 0.5-1.2 litre; 1.2 litre and larger sizes), paying the recycler an amount of 5, 10 and 15 *fen* Renminbi respectively, the equivalent of one half to 2 Eurocents (January 2017 rates). One can also refuse payment and simply donate one's recyclables (Su, 2014).

To operate RVMs in an urban Chinese environment, often outdoors, the machines had to be redesigned to meet specific requirements. To this end, Incom received financial support from the National Development and Reform Commission under the State Council to develop the machines currently operating (Kanthor, 2015). Incom rolled out its first RVMs for PET bottles in selected subway stations along Beijing's No. 10 subway line in 2012 (Zhang and Wen, 2014). The pilot project aimed to have 3000 RVMs in operation in the city, not only in subway stations, but also near or in schools, residential areas, bus stops, and shopping malls (Illustration 2.2). Once the trial phase had ended successfully, the company planned to have 10,000 units working in Beijing and 14 other large cities, including Shanghai and Shenzhen in 2016.

**Illustration 2.2   Two RVMs operated by the Beijing Incom (Yingchuang) Resources
Recovery Recycling Co. Ltd, at the Panjiayuan No. 10 Subway
Station**

Author's photograph, 9 April 2017

The RVMs used for PET collection are integrated with machine-to-machine
(M2M) communication technology. Once an RVM unit is full, it sends a message
to the company's depot in Beijing's Shunyi district, whereupon a truck is
dispatched to empty it. Its contents are then sorted and recycled in government-
approved facilities and remade into new PET bottles (Kanthor, 2015). On the
basis of fieldwork observations, I have established that this retrieving does not
take place as often as the company maintains; for consecutive days, many of
the RVMs that can be found on the company's website map were completely
full without being emptied (marked in red) (Fieldnotes, 2017).[9] The big data
that are collected through the M2M-process make the system attractive to the
big brands producing beverages as well as to the government. The machines
collect information from scanning the bar codes on the bottles and can provide
a picture of current trends in consumption and recycling. Until now, Incom has
supplied these data free-of-charge to the government and the brands that buy
advertising space on the surfaces of the RVMs. Given the expertise it has gained
in the process, the company has become one of the largest manufacturers
and exporters of solid waste RVMs in China, but is also finding customers
for its machines in Mexico, the Netherlands, Brazil, Thailand, and elsewhere.

9   A map of Incom's RVM locations in Beijing can be found here: http://www.incomportal.
com/baidu/baidu.jsp.

One of the main problems encountered by Incom has been the fact that the supply of raw waste materials, i.e., used PET bottles, for recycling is fluctuating, and that the net yield per RVM cannot be predicted. One can safely assume that beverage consumption is higher in periods of warm weather than in colder times, but aside from seasonal influences it is impossible to estimate the number of bottles that will be returned. Practice has shown that the amount of recyclable PET obtained from the RVMs remains far behind projected yields. Consequently, the recycling facilities that Incom has imported and installed cannot run at their full capacity. According to Incom, the reason is that the competition between RVMs and individual scrap collectors, the people roaming the streets, is too fierce. Or, to put it differently, it is more convenient for consumers disposing of empty PET bottles to dump them wherever and whenever they feel like it, in a waste bin or simply on the street, instead of making a conscious effort to go find an RVM. Other O2O recycling platforms encounter similar problems in obtaining recyclables (Guo, 2016).

To alleviate this problem of contestation over access to recyclables, Incom has taken the opportunities offered by the government's Internet Plus policies a step further by developing an online-to-offline strategy to supplement the supply and demand of recyclable materials. Incom has added the smartphones of individual consumers or residential communities as a link in this chain. In May 2015, the company launched a smartphone application called Bangdaojia ('Help at Home') that can be installed and used by the consumer; the app runs on a host of mobile platforms, i.e., as a stand-alone app on Apple or Android, or integrated into Tencent's Weixin/Wechat. By 2017, the service covered 400 communities in Beijing (Zheng, 2017). Incom is not the first company to develop an app, but judging by the frequent media write-ups, it is certainly one of the most successful in drawing the attention of both Chinese and foreign media (PressTV Reporter's File, 2014; Kanthor, 2015; Zhang, 2016). This has made the solution adopted by Incom into a national (and potentially global) model for others to follow.

The Bangdaojia app can be downloaded by clicking the relevant link on the Incom website,[10] which transports the consumer to the app page, and scanning the QR-code placed there.[11] Once downloaded on a smartphone and registered by using the consumer's mobile phone number, anybody who has waste paper, plastic bottles, or other waste materials waiting to be picked up can report it in the app; at the same time, the app calculates how

10   http://www.incomrecycle.com/
11   http://www.365bdj.com/

much the scrap offered for sale is worth. This option is not only available to individual consumers; the Bangdaojia website makes clear that the app can be used for/by residential communities, schools, and other (commercial) entities, ranging from offices to restaurants and shops. After indicating that scrap can be picked up, one of the 300 waste collectors employed by the company contacts the person that is offering the scrap, picks it up, and then transfers the money through a wallet app (Zheng, 2017). It remains unclear how Incom has hired the waste collectors, and how their contracts are structured. On their page of job offerings, analysed in Chapter 4, there are no vacancies for waste collectors listed.

Bangdaojia also gives access to a map that tells the consumer where the nearest RVM for PET bottles is located. By summer 2017, the app saw the addition of tabs for locations for the disposal of broken-down smartphones and used clothing, further proof that Incom is expanding its activities. By that time, the rejected smartphone market had already been well established and divided. The textile drop-off points was rather small in number and widely scattered, which does not bode well for the initiative. Moreover, the market for reused and recycled clothing in China is inefficient and riddled with scandals and corruption (Ma, 2017). However, all these steps contribute to the start of a recycling retrieval model that includes all types of scrap that are on offer.

Effectiveness of the O2O model

It is hard to say whether these companies have been successful with their recyclable retrieval activities and/or maintenance jobs, or whether they will continue to be so once the novelty of their operations has worn off. One of the main problems they encounter is the question of scale. In 2017, their services were not offered on a city-wide basis but were instead concentrated in the Eastern districts, i.e., Dongcheng and Chaoyang. Some had a few small bases of operations in the Xicheng and Fengtai districts, but these did not contribute to the bulk of their activities. The RVMs operated by Incom are the only ones that can claim citywide coverage. They are distributed fairly evenly over the breadth of the city, reaching out into even the suburban districts of Shijingshan, Changping, and Huairou; however, they are not encountered in subway stations as often as the company wants us to believe. Zhang and Wen (2014: 997) established in their research that their respondents were quite taken with the idea of the RVMs that Incom operates and were convinced that they would be useful for the collection of recyclable bottles.

Illustration 2.3   A van operated by Taoqibao, at a Qianbajia waste collection point, formerly part of Henan Village, Haidian District, Beijing

Author's photograph, 29 March 2017

Despite the amount of publicity given to their presence and write-ups about O2O activities in the (local and foreign language) media, and despite the various tools and machines that are conspicuously shown in these narratives, I did not spot a single O2O-company retrieval truck during the three months of spring 2017. I only saw two vanlets operated by Taoqibao at one single garbage collection point in the Qianbajia area in Haidian district, the original Henan Village (Illustration 2.3). This part of town is actually not serviced by Taoqibao at all, as the company operated out of offices in Xicheng. Moreover, these vans gave a rather suspicious impression. Although they sported the Taoqibao logo and colours and carried the slogans the company used, they had obviously been painted over recently and had previously served as delivery vans for another company. This made both me and a number of bystanders wonder whether these vanlets were actually operated by Taoqibao, or if they had been *shanzhai*-ed, i.e., copied by others attempting to operate under the guise of the Taoqibao company for their own purposes. If it was worthwhile to copy the practices of Taoqibao, it must mean that the company was hugely successful. As there are questions surrounding

Taoqibao's success, it seems that some (informal?) waste pickers may have thought that having recognizable company colours would improve their income. The remarks and opinions of waste collectors and collection point bosses that I recorded and that I present in Chapter 4 suggest otherwise.

As my interviews showed, not many of the intended users were actually familiar with the services the O2O-companies offered. Most people had never heard of them; others, without actually knowing anything about them or their business models, immediately characterized them as only being interested in valuable waste, in particular e-waste. Various respondents acknowledged that their services could be convenient but reckoned that they were not really needed. They were elated that the set prices for recyclables in the applications meant that waste pickers would no longer be able to cheat. And they hoped that the applications would bring more order to the waste-collecting business, meaning that their communal waste bins would no longer be ransacked by informal scavengers, with the waste strewn about carelessly. However, none of these considerations made them contemplate joining an O2O scheme.

3   The human factor – garbage producers

This chapter concentrates on the residents of Beijing who produce garbage every day. On the basis of interviews and field data gathered in spring 2017, I focus on the following questions. Where do Beijingers stand when it comes to dealing with the waste they produce? Are they aware that their increased consumption exacerbates the problems related to its disposal? Are they inclined to embrace the circular economy and start to reduce, reuse, and recycle? What do they think about garbage retrieval services like the new O2O-companies introduced in the previous chapter?

Residential communities

At the risk of oversimplifying, Chinese urbanites tend to reside in *xiao qu* ('residential communities') of approximately 2000 households each. A number of these residential communities are overseen by residents' committees, which in turn are grouped together into *shequ* ('neighbourhood communities'). These fall under the jurisdiction of street affairs offices, which fall under one of the sixteen urban or suburban *qu* (district governments) (Ngeow, 2012; Teets, 2013: 8; Hu, Tu, and Wu, 2018). Large parts of the housing stock that presently make up the residences of a residential community were originally distributed and managed by the *danwei* ('work unit') (Gu, 2001: 95). During the Maoist era, these work units together served as a 'mini-welfare state' (Gu, 2001: 91) for urban workers, as they were responsible for 'providing employment; collecting taxes; allocating welfare resources; monitoring and controlling the movement of employees; certifying births, deaths, and marriages; and implementing government policies' (Ngeow, 2012: 77). The *danwei* system made urban workers highly dependent on their units and, by extension, the state (Dutton, 1998: 42-61; Gu, 2001: 92). By the 1990s, financial constraints had forced the work units to relinquish most of their welfare functions, including housing, medical care, education, childcare, etc. The housing stock that had been at the disposal of the work units was sold off, often for a discount, to the people inhabiting it (Davis, 2005; Yang and Shen, 2008; Hu, Tu, and Wu, 2018).

Even after urban construction took off in Beijing in the 1980s, the former work units continued to dominate the housing market. They negotiated preferential prices for their employees when new housing projects were constructed for sale. Many inhabitants of residential communities feel a sense of pride and belonging, based on the fact that they were among the original occupiers of an apartment owned or sold by their work unit. In the eyes of many, attachment to a state-owned enterprise, however virtual this link may seem now, elevates one's position and raises one's social capital. Over the intervening years, many of the original inhabitants of the residential communities have moved on to larger apartments or to other cities and have sold or sub-let their original place of living (Hu, Tu, and Wu, 2018). This has led to a situation where the original inhabitants who continue to live in the community see themselves as having a higher quality, or *suzhi*, than the newcomers. The importance of this concept of 'quality' is discussed in more detail in Chapter 5. The 'new' residents have no shared history or experiences with the first generation of inhabitants and have no ties to the original community, as they moved in comparatively recently.

Housing has become a marker of consumerism and thus of status. The neighbourhood and location of one's dwelling confers status. Most commonly encountered are: commercial housing communities, gentrified and often gated, that have higher status; and old or after-sale public housing communities, i.e., housing that has been sold off by state-owned enterprises to former employees. There are also hybrid forms (Hu, Tu, and Wu, 2018). Even more important, although less visible, than the location of the apartment is the way in which it is furnished (Fraser, 2000; Davis, 2005). As the government's attention shifted away from almost exclusive concern for ideological conformity to economic development under the Reform policies of the 1980s, the act of consuming came to be inscribed with political meaning. Buying things supported the development of the economy; buying more not only served to demonstrate support for the new policies, but also signified a kind of distinction, replacing other erstwhile ideological and political markers such as class background (Li, 2010: 153).

Municipal classification and separation pilots

Since 2000, the Beijing municipal government has tried to get its citizens to classify and separate their garbage. Many other urban centres in China (Shanghai, Hangzhou, Suzhou, and Shenyang, to name just a few) have also attempted the same, using a variety of approaches. The Beijing government has resorted to using a number of methods, many introduced on a trial basis in designated residential communities. It has installed separate garbage bins for different categories of garbage for household use; it has made colour-coded garbage bags available to make sorting easier; it has

Illustration 3.1   Communal garbage bins on the grounds of a residential community. The sign indicates which garbage needs to go into which bin

Author's photograph, May 2017

experimented with user fees on the basis of the principle that the polluter pays; and so on (Yuan et al., 2016) (Illustration 3.1). Each of these methods has had some successes as well as failures, and none has made it beyond the pilot stage. The number of residential communities where garbage classification regimes were practiced increased from 600 in 2010 to more than 2400 in 2012 (Friends of Nature, 2013; Yang, 2013; Yuan and Yabe, 2014). I was surprised to find that many residents of such pilot communities had not been informed of the fact that they were taking part in an experiment. They had no idea that their behaviour was part of higher government plans and as a result were not familiar with the aims and targets of the movement. This had a negative effect on their motivation to seriously participate.

In all of these experimental cases, the municipal administration shifted the burden of the implementation to the street affairs offices' levels, who have shifted it to community levels, who have shifted it to, finally, the community management offices that are responsible for the actual community. No matter which method has been attempted in whichever urban centre or district, the citizens have not made any commitment to embrace any of the schemes, citing various reasons why they still do not classify and recycle their waste. These range from lacking space in their apartments to the fact that recycling activities were too cumbersome or confusing. When trying to determine the reasons for this tendency to not participate fully,

Illustration 3.2   A signboard put up in a residential community providing
information on garbage classification and separation. The
message stuck on the board urges people to dispose of their
garbage in the bins, in order to promote civilization and health

Author's photograph, May 2017

Yuan et al. (2016) have concluded that active citizens' resistance, political complexities, and cultural origins are behind the lack of success when it comes to fostering structural, long-lasting recycling behaviour. As Simões has argued, 'citizens are not necessarily willing to be informed and – more importantly – to act properly upon that information. Individuals are free to refuse such information, to ignore it, or not to change their behaviour' (2016, no page). According to some of my sources who have been keeping track of the various (municipal) campaigns to make garbage separation an inalienable part of daily life, local governments have not performed well in the disposal of garbage in general. After all, it does not make any sense to separate one's garbage when the sanitation department is not equipped to dispose of the garbage adequately or when separated waste treatment is not in place in general (Martens, 2006). Some of the trial projects ended up as merely face-saving efforts, with the responsible departments only going through the motions of taking action. In the case of Beijing, for example, some districts enlarged the garbage bins in their areas or provided garbage bins with openings for separate types of waste, but with only a single receptacle under the lid, turning the separation activity into a farce; or they installed an embryonic information collecting system that quickly sputtered out; none of these measures actually succeeded in changing the habits or behaviour of the residents (PKU MBA Deep Dive, 2015; Interview with Hong Chao, 2017) (Illustration 3.2).

Gender and garbage

At this point, it is relevant to point out that the practices involved in garbage disposal, classification, and/or separation in the urban configurations I studied in Beijing are gendered activities. It is usually the women in the household who are responsible for the garbage and this situation certainly is not unique for Beijing (Li S., 2002: 794-796). In the interviews and informal talks I held with community residents, the female respondents often indicated that they were the ones who actually dirtied their hands doing the garbage. Based on the manner in which they talked about these topics, it was clear that this was completely natural in their opinion, as it was part of how they visualized their housekeeping work. Many of the housewives expressed a desire to take a more active part in waste reduction and/or garbage classification and separation, thus offering more of a personal contribution to improving the environment for the next generation. But they also let it be known that they simply did not know how, were not told what to do beyond what they already knew about, i.e., selling the most valuable recyclables (Zhang and Wen, 2014; Interviews, 2017).

Their spouses or the other male residents present were quick to explain that they were just as involved in keeping the house clean of garbage as their wives were. They insisted quite vocally that taking care of the garbage was a man's job, and therefore they saw it as their responsibility. The wives often furtively made clear that this was just empty talk. Some male residents specifically mentioned the fact that garbage disposal (but not separation) was something that all household members, including the children, participated in. These fathers suggested that the matter of garbage was a subject they discussed often in the household. It offered them ample opportunities to teach the younger members of the family about values and social responsibilities. Indeed, the male members of the household tended to be quite familiar with and supportive of the current phrases, slogans, and concepts that were used by the government or community management offices when I talked with them; they quoted them often and literally (Interviews, 2017).

Some of the children who were around during the interviews with their parents did not take part in classifying or recycling at all. As members of the One Child and post-1980s or -1990s generation, they usually were not asked to do anything in the household but were instead told to concentrate on their studies; the future welfare, wellbeing, and status of the family depended on them alone (Fong, 2004). The small number of kids who occasionally were ordered to dispose of the garbage made no secret of the fact that they loathed it; most said that this was something Mom usually did, or should be doing, not their job. Similarly, they had no clear ideas about the details of garbage disposal or how to classify and separate waste; these activities were simply not within their sphere of interest and they could not be bothered about them, simply dumping their waste wherever was convenient (Interviews, 2017).

Age and garbage

In talks, interviews, and discussions, elderly citizens tended to complain about the fact that the younger generation did not care a bit about recycling. The older people were quick to characterize children as a throw-away generation, disposing of the waste they generated without a second thought. Their own traditional values of communist austerity where it concerns the use of things were overturned by the values of consumerist hedonism embraced by the generations that were born after the 1970s (Wang, 2002; Li S., 2002; Simões, 2016). As these older generations fade away, their recycling values and knowledge rooted in practical economic and environmental awareness are also disappearing (Williams, 2014: 296). The idea that the elderly are more conscious of separating and recycling also was reflected in the educational materials produced by the Solid Waste Management Office of the Beijing Municipal Appearance Management Committee, which are analysed and discussed in more detail in Chapter 5. These publications present the grandmothers – and only occasionally grandfathers – as the ones who are fully aware of what to do with the garbage and how to sort and separate it, as opposed to the other members of the family. In popular perception, the elderly do not need to be told or educated.

Retirement starts early in China, although steps are underway to raise retirement age. Urban women working for a state enterprise or private company can retire when they are 50 years old, while men usually work until they are 60; civil servants generally work five years longer than enterprise employees. For urban workers, 54 years is the average age of retirement. Upon retiring, urban workers can expect to receive about 2000 yuan per month. This is in stark contrast with farmers, who receive only 70 yuan per month once they have passed their 60th birthday, if they are lucky (Shi et al., 2015). Pensions are just another part of the basic provisions that are bound up with the *hukou* ('Household Registration System'), which favours urbanites over the rest of the population (discussed in more detail in Chapter 4). Migrant workers in urban areas can claim a pension, but only in the place they have been registered.

Many of the elderly people one encounters in town often give an impression of having a carefree existence; still, they make a point of collecting recyclables wherever and whenever they can. One rarely spots an old person without a shopping net or shopping cart, often filled with empty plastic bottles. Well-dressed pensioners, usually women, sorting through trash cans is a sight frequently seen. Many insist that this sort of activity keeps them spry and energetic, as if recycling is part of some self-designed fitness programme. They also make clear that they see their hunt for recyclables as a civic duty, and as part of their *suzhi* ('quality'). Many talk with disdain about young people in general, and express their disgust with the ease with which the latter consume and simply throw stuff away, preferably making an effort not to put their trash into the waste bins. The fact that the elderly show a higher awareness of wastefulness and recycling should not come as a surprise. Many of them have lived through times of hardship and want; many have witnessed periods of famine and destitution (Evans, 2014; Wang et al., 2016); many were thoroughly socialized in the 1950s and later to collect junk for the state, with the result that they see recycling as an essential activity (Li S., 2002).

Most of my elderly conversation partners scoffed at my suggestion that their pensions were inadequate and that they needed the extra income they could earn by selling used bottles or other valuable junk. They very carefully explained that they were well taken care of by the government. Rather, they said, collecting recyclables is like finding free money on the street that can be spent on the grandchildren or to do something special for themselves (Li S., 2002; Goh, 2009; Griffiths, 2014; McKinsey & Company and Ocean Conservancy, 2015; Interviews, 2017). In opposition to these apparently carefree retirees, there are large numbers of elderly people, both male and female, who do not create the impression that they dig through trashcans for fun. This is not limited to specific parts of the city, but they are generally more active in the poorer sections and neighbourhoods. It is often hard to distinguish whether these individuals are part of the growing numbers of the urban underclass, made up of former state-owned enterprise workers who have been made redundant, or elderly migrant workers (Zhang, 2002; Solinger, 2004; Liu and Wu, 2006; Saunders and Sun, 2006; Gao and Zhai, 2017). Many depend on recycling PET bottles and other items of recyclable garbage simply to survive; with no or inadequate pensions to fall back on, their income derives from whatever they can earn from sorting, collecting, and selling waste (Solinger, 2006; Taylor, 2008; Griffiths, 2014; McKinsey & Company and Ocean Conservancy, 2015; Gao and Zhai, 2017; Fieldnotes, 2015, 2017; Interviews, 2017).

One false move

Recycling initiatives tend to fail when the residents are insufficiently convinced of the effectiveness of their separating behaviour, whether this is newly acquired or deeply ingrained. One experience that I heard recounted often, told by people living in different communities as well as environmental NGO representatives, is how garbage that the residents had carefully classified and separated ended up in one single individual all-purpose garbage truck and got mixed with the other unsorted garbage (Friends of Nature, 2013; Interviews, 2017). Talk about such incidents, even if they occur only one time, tends to continue forever. People return to the topic over and over again. And it makes those willing to classify and separate start to wonder why they should bother at all, when in the end it all ends up in the same truck, the same landfill, or the same incinerator anyway (Martens, 2006). Research has shown that in many communities, no information was provided, for example on notice boards, about where the carefully classified and separated garbage would end up. This had a negative effect on the possibility of motivating the residents (Friends of Nature, 2013)

Despite this, further research has shown that 59 percent of Beijing's residents said that they were aware of the seriousness of the city's garbage problem. The ENGO Friends of Nature discovered in 2012 that 71 percent of the residents reported that they classified and separated garbage at home, while only 63 percent did so in 2010. All the same, many admitted that they had never heard of garbage classification and separation, and multiple reasons stopped people from doing so. Some said that they lacked awareness and could not adopt the habit; some thought it should be the responsibility of the real estate company or the community management committee, as they had already paid them fees. Some said they could not be bothered, especially after a heavy day's work. Some found it unnecessary; others felt less motivated when fellow residents did not do the same (Friends of Nature, 2013).

This corresponds well with the results of my own interviews. Many people insisted that they duly classified and separated their garbage into recyclables, non-recyclables, and plastics. As Tang, Chen, and Luo have established, however, 'self-reported recycling behaviour and recycling behaviour itself are not tantamount' (2011: 860). Other interviewees, often those who lived in communities that had been designated as experimental bases for waste separation by the municipal government, indicated that they even went beyond what was required and had started to separate their kitchen waste as well. In some communities, green kitchen waste containers had been introduced and installed with clear usage instructions (Friends of Nature, 2013; Yuan et al., 2016). Practically all of the interviewees had heard or read stories, online or in newspapers, or had seen TV programmes that reported on the more advanced and sophisticated waste separation practices in Japan, Taiwan, and elsewhere (Ding, 2016; Huishouge, 2017a). Although they admired these examples and considered them interesting and potentially relevant, the Beijingers still saw them as 'foreign' – and this was usually the most explicit reason given for not following them. Aside from offering arguments that such practices were not Chinese, residents in general said the lack of space in their apartments was one of the main reasons they did not classify their garbage into more narrowly defined categories. This did not stop them, however, from using hallways and other shared spaces to store goods they did not (seem to) use any longer (Fieldnotes, 2015, 2017; Interviews, 2017). The knowledge needed for garbage separation into the detailed categories that are in use in Japan and Taiwan is completely non-existent among Beijing community residents, particularly when set against their general lack of understanding about the need and urgency to start doing so. They attributed their lack of awareness to the fact there are no structured or continuous propaganda or educational programmes or information drives about these topics. While Minter (2013a, 2013b, 2015) and others (Li S., 2002: 788-794; Interviews, 2017) convincingly argue that Chinese are very knowledgeable about the potential sales value of their waste, other sources stressed that Chinese consumers have no clue that their waste is worth something – that, for example, white plastic recyclables sell for a different price than black or coloured plastics, or that transparent glass is re-used differently than coloured glass. The sources who stressed the need to make residents aware of the value of recyclables were convinced that once they became aware of the opportunity to earn money, more residents would be willing to start recycling (Interview with Hong Chao, 2016).

Illustration 3.3   Coloured communal garbage bins on the grounds of a Haidian
District residential community

Author's photograph, April 2017

Most residents live in communities that have three communal garbage containers (one for recyclables, one for non-recyclables, and one for plastics) in the yard, or combinations of this configuration (Illustration 3.3). They consider it less cumbersome to hop down the elevators several times a day to deposit their waste in the container they think is designated for their specific type of waste than to keep the garbage indoors in some special receptacle for future disposal. Those who had not taken part in any of the experiments related to kitchen waste were convinced that having to store food waste indoors would surely be a nuisance, particularly in summer when it quickly developed noxious smells and attracted all sorts of unwanted vermin. They were not sure whether they would ever start separating their kitchen waste for this reason (Interviews, 2017).

Many of the middle-class residents confessed that although they had classified and separated their garbage before, they had stopped doing so for various reasons. This change in behaviour most certainly was not the result of a sudden decrease of 'quality' that they might have suffered. On the contrary, almost all of them prided themselves on their high levels of environmental awareness, and many mentioned that they had taken an active part in projects that had been organized in their communities and/or by ENGOs. They were generally outspoken about the need to classify and recycle and expressed their desire to help make the world a better place to live, both for themselves and for their children. But still they had given up separating their garbage, often after a rather short period of time. Since they did not feel good about their actions, they said that others were responsible for making them stop, or at least for not supporting their classification and separation efforts. Those 'others' had persuaded them that it all cost too much energy and involved too much trouble. This happened in particular in communities where there was a 'guy downstairs' who collected, separated, and sold the community's garbage for a living. Why then should they bother to spend energy on these activities (Interviews, 2017)?

The 'guy downstairs'

This introduces the phenomenon and importance of the 'guy downstairs' into the discussion. The 'guy downstairs' is a fixture in the chain of urban recycling that has evolved in Beijing and other urban centres. These 'guys downstairs' are more often than not actually husband-and-wife teams, members of the army of migrant labourers from the countryside. They may have started out as waste pickers roaming the streets, scouring waste bins, and hunting for recyclables. In one way or another, they have had the opportunity to formalize these activities by offering their services to a residential community while still maintaining their independence and remaining part of the informal sector of waste collectors. They often have been invited by the community management office or real estate agency to take care of the solid recyclable waste without being offered formal employment. Others have 'bought' the right to collect the recyclables of 'their' specific residential community (Ensmenger, Goldstein, and Mack, 2005; China Youth News, 2016; Interviews, 2017). Over time, some have become such fixtures in the communities where they ply their trade that they consider them to be their actual posts. In many communities, these waste collectors actually occupy a portacabin-like cubicle from which they operate and where they store recyclables awaiting transportation upstream (Beijing Municipal Urban Management Committee and Beijing Municipal Urban Management Committee Information Center, 2017: 2). Their scrupulous sorting of garbage complements the hauling work done by the city garbage collectors who come and empty the bins in the yard. The 'guys downstairs' have a negotiated yet informal base of operations and generally are responsible for maintaining the yard; in short, their job is to keep the community in tiptop shape. Their income is based on whatever price they are able to negotiate for the recyclables they have collected (Interviews, 2017).

In terms of work conditions, having a semi-permanent, semi-formal job within a community is seen by many migrant workers as a step up, as a symbolic improvement of their lives, almost like having steady work while still remaining employed informally, with all the connotations of being an independent operator (Ezeah, Fazakerley, and Roberts, 2013). Given its semi-formal characteristics, this type of employment does not qualify the 'guy downstairs' to apply for an urban registration. As informal garbage workers, they are considered inferior; without an urban registration, they can be nothing but sojourners, of the place, but not from the place. This stands in stark contrast with the attitudes that reportedly prevailed in Beijing during the Republican era (Goldstein, 2006). Given the possibilities of rich yields of recyclables, these in-yard posts are much sought after. They could serve as the starting point to develop more structural sustainable community activities, depending on whether a migrant worker or potential 'guy downstairs', is willing and able to take initiatives (Tong and Tao, 2016).

Not many respondents had personal contacts with 'guys downstairs', nor were they particularly cordial or close. Their interactions were fleeting, not going beyond an occasional greeting every once in a while, or a casual 'here you are' remark when handing over a piece of recyclable garbage. Many residents did mention that they put their recyclables not inside the garbage bins but on top of or next to them, making it easier for the 'guy downstairs' to separate and collect. These 'guys downstairs' simply were there, human beings that community members did not need to acknowledge. In many cases, the residents did not register their presence at all, or suggested that they had already left the community to go to work by the time the 'guys downstairs' entered. In a few instances, respondents actually said that these people were too uncultured (in the sense of lacking 'quality'), too dirty, too malodorous to communicate with; others hinted that they could not be trusted, that they had criminal leanings (Mobrand, 2006; Tse, 2016; Interviews, 2017). In short, the people working in the yards collecting garbage were tolerated. Many suggested that elderly residents might have closer relations with these 'guys downstairs' or strike up friendships with them, often looking out for them and being solicitous about their wellbeing. But this was attributed to the fact that the elderly were always looking for some company, somebody to talk to in order to while away the time. And the people working in the yards were also ideal sources of gossip, as they observed the daily comings and goings of the residents.

Case study of Red Nest Community Resource Centres

The phenomenon of the 'guy downstairs' has gone through an O2O metamorphosis under the Internet Plus Recycle plan, similar to the transformation of the door-to-door recyclable collection process. In cooperation with the Synergy and New Ecology Design Centre of Tsinghua University and the Urban Development and Environment Research Centre under the Chinese Academy of Social Sciences, the Hong Chao (Red Nest) Enviro-Tech Company has developed an approach that modernizes the collection activities undertaken by persons in the yard. The aim of this solution, which combines O2O and M2M (machine-to-machine) technology, is to extract big data from collection activities; to streamline the collection, disposal, and recycling of waste materials; and to provide continuous educational moments and awareness-raising opportunities among the community residents. This approach has been chosen on the basis of extensive research in communities in Hebei Province and Beijing proper that was undertaken by the three units involved. One of the outcomes of the preliminary research was that the majority of respondents were highly motivated to engage in garbage separation and recycling, but unable to do so for several reasons: first, they lacked knowledge and understanding of the process; second, the governments at various levels had failed to make clear why such behavioural change was necessary; and third, the respondents felt that the existing garbage disposal system was unsatisfactory and lead to undesired secondary pollution. This secondary pollution was partly the result of the too-infrequent visits of garbage trucks coming to empty the communal bins, partly of waste pickers' continuous rummaging through the often-overflowing bins. In Spring 2017, the Hong Chao company had been running a number of pilot sites in selected residential communities for some 15 months. Some of these pilot sites were designated after prolonged negotiations with local governments; some joined the scheme voluntarily. The Hong Chao company entertains the idea that in the future, after a prolonged period of testing and community presence, it will be able to play a role in providing the government with the data needed to formulate policies that aim to make the user pay for the garbage s/he disposes of (Interview with Hong Chao, 2017).

The forms and functions of these Red Nests, termed 'Community Resource Centres', are not altogether new. In various residential communities, the waste collectors downstairs already operate from a small cubicle of varying quality and comfort (Tong and Tao, 2016; Fieldnotes, 2015, 2017; Beijing Municipal Urban Management Committee and Beijing Municipal Urban Management Committee Information Center, 2017: 2). Red Nests

**Illustration 3.4   Front view of a Red Nest in a Shunyi District residential
community, designed and operated by the Hong Chao (Red Nest)
Enviro-Tech Company. Note the RFID tag dispenser**

Author's photograph, 4 May 2017

essentially have the same characteristics as the portable buildings, or
portacabins, which are already in use in many communities. This makes
them ideally suited for introduction to pre-existing built environments
and communities where no such services had been offered before: they
may be new, but in essence they are not. Despite the pre-fabricated and
familiar appearance, it is obvious that a serious amount of design has been
involved in the way the Red Nest structure looks and the many functions
it has. The Nests are in fact small collection stations. The facade of the
Nest is fronted by two sliding windows with four hoppers underneath,
each clearly identified as receptacles for valuable, dry, wet, and poisonous
waste by means of stickers with cartoon symbols. These symbols have
been designed in such a way that even illiterate people can understand
what type of waste should go where. Waste that is too bulky to fit in the
hoppers can be delivered through the windows or the door. Slightly to the
right and stretching the full height of the structure is a column for the
disposal of batteries and e-waste. To the right, a video screen broadcasts
looping messages about separation, recycling, and other related topics. On
the roof of the structure a battery of solar energy collectors is mounted,
providing the whole setup with the energy required to operate. A ticket
dispenser is mounted above the hoppers, which is activated through a
proprietary app that community residents download on their smartphones.

The dispenser produces RFID-tags that need to be stuck to the garbage that is disposed of in the hopper (Hongchao Enviro-Tech, n.d.; Fieldnotes 2017) (Illustration 3.4).

These RFID-tags are essential for the collection of the big data that will drive the Red Nest network once it has grown, for it is within the portable building that the actual second separation takes place. The interior is dominated by a sorting table in the middle, underneath the sliding windows, with an RFID reader hanging over it. An employee empties the hoppers, scans the RFID tag attached to the waste, sorts it, and disposes of it in one of the numerous designated cabinets that are on the side. These cabinets, which can be accessed from both within and outside the structure, have sensors that keep track of volume and/or weight. Once the cabinets are filled, the system sends a notification to the recycling companies that a certain kind of garbage container is full and ready for pick-up. The recycling companies pay the Red Nest company and move the waste up the stream towards recycling or ultimate disposal (Interview with Hong Chao, 2017).

While the Red Nests serve as collection points where people can drop off their waste, this is not their only or even main function. The proprietary app that the company has developed allows community residents to notify the person(s) manning the Nest that there is garbage to be picked up at their doorsteps. In ways that are similar to the O2O companies discussed in Chapter 2, the residents who offer waste for pick-up collect credits that they can use to trade for goods. The garbage thus collected is treated in the same way as the waste delivered to the Nest cubicle proper: it is RFID-tagged and sorted for a second time, thus adding to the body of data that Hong Chao wants to amass. Both the persons offering the garbage and those collecting it should feel more assured: the residents are certain that their waste is dealt with correctly and speedily without having to turn to or wait for outsiders, and the waste retrievers can collect waste more efficiently (Interview with Hong Chao, 2017). Moreover, a system like this is transparent and does not leave room for cheating by waste buyers. It scares off the scavengers who roam communities, ripping open plastic garbage bags and spilling their contents. These were complaints that were voiced in other sources as well as by interviewees (Goldstein, 2006; Interviews, 2017) (Illustration 3.5).

Indeed, one of the most striking characteristics of the Red Nest plan is how it integrates waste separation and disposal within the community and makes it completely visible to all concerned. All interactions between residents and waste workers are regulated and take place in public. This

Illustration 3.5   A residential courtyard in Haidian District, with three communal bins and a collecting bin for textiles

Author's photograph, May 2017

allows for a familiarity to develop between residents and waste workers, which in turn opens up opportunities for continuous education that goes beyond the pasting of slogans or themed posters, or even the materials that are broadcast in endless loops on the video screens of the Red Nests themselves. The regular interactions with the waste worker stationed in the Nest may turn him/her into something of an acquaintance, whose urging to separate and recycle may more easily be taken to heart. Even though they work within a community, the people employed in the Red Nests will never become full members of that community. Having a stationary work spot, however, allows them to work more efficiently, in a better environment, with a fixed income and social benefits (Interview with Hong Chao, 2017).

Using O2O apps and services

The O2O companies obviously face stiff competition from initiatives like the Hong Chao solution, because the latter seems much more similar to the existing 'guy downstairs'. Moreover, many of my interviewees indicated that the O2O services were completely unknown to them, the potential users. When I explained what the aims of companies like Zaishenghuo, Bangdaojia, and Taoqibao were, how they operated, and how their smartphone apps worked, most residents made it clear that they had never heard of these companies or what they offered. They had never seen one of their workers or vans in the streets or in the communities. None of their relatives living nearby or in other parts of the city had mentioned them even once. When asked, some said that they were familiar with the Incom RVMs in the subway stations, maybe had even used them once or twice and found them convenient, but they stated that they would not go out of their way to find one to dispose of a PET bottle. They were not aware that Incom also ran the Bangdaojia service.

My interviewees were not attracted to the potential convenience the apps offered, though they admitted that having service people coming to one's door for small repairs and/or odd jobs did sound very convenient and appealing. According to many respondents, downloading and using an app like this was something that only younger people did, not something for them. This immediately turned the question about whether they would join an O2O scheme into a statement that it might attract the younger, higher salaried generations, but it was not something that ordinary people (*laobaixing*, the 'old one hundred names' is the term my sources would use) would think of doing. They suggested that they considered recycling as something of a daily responsibility that they took care of themselves. They continued to say that the elderly either did not own a smartphone or were not able to operate them. This view was not supported by my own experiences or observations (Fieldnotes, 2015, 2017; Interviews, 2017). Many Chinese media reports show that the elderly are considered to be a particularly promising target group for O2O companies (Zheng J., 2017).

Many interviewees contended that they did not need a service in their communities to pick up recyclables on their doorstep, as they already had 'guys downstairs' who took care of the recycling. Moreover, they were convinced that these companies were only interested in specific types of recyclables – valuable ones. They did not perceive any potential of convenience from O2O services; instead, they expressed dismay at the idea of having to notify various different companies to help them dispose of their assorted junk. Others were surprised at the various maintenance services offered by the companies. They wondered who would hire such a service: according to them, these things should actually be taken care of by the community management committees, and as residents they had already paid a fee for such work. Why would they pay extra money? Most of my sources also feared that O2O services would be picky and difficult to work with, as they were only interested in valuable junk like used smartphones, and not the sort of trash that my interviewees would usually have on offer. They seriously wondered why they would ever engage such a service (Interviews, 2017).

Many respondents referred often to worries about what to do with the large numbers of used batteries they ended up with. They were aware of and concerned about the effects on overall pollution that batteries would have, something that must have stuck in their memories after watching television programmes or reading articles in the media. They were generally anxious to dispose of batteries properly, much more than the other waste they generated. This may point to the positive effects that sustained publicity about the negative effects of some goods can have on consumer behaviour. But the anxiety of the respondents made it clear that, while they discussed this problem amongst themselves, no solutions were offered by the management committees or district or city governments to remedy the situation, and no action was undertaken, such as providing special disposal bins for used batteries at strategic spots. When residents suggested solutions to their management committees, they felt that their suggestions were not taken seriously (Interviews, 2017).

Ultra-*suzhi*

Some of my scholarly colleagues and friends, though admittedly not a large number, have made a start with their own type of civic action concerning waste disposal. Whenever they see an empty bottle, box, or discarded newspaper lying in the street, they pick it up and dispose of it in the nearest garbage bin. Driven by a sense of propriety, they attempt to educate others by setting an example. In the process, they demonstrate that they have an ultra-high level of *suzhi*. Many times I have seen astonished and non-comprehending onlookers witness such actions; aside from expressing their surprise, however, they did not seem to be educated or influenced in any way by this exemplary behaviour (Fieldnotes, 2015, 2017).

It is remarkable how this proactive behaviour corresponds with similar phenomena in some Western countries. Well-educated members of the middle classes there have taken to voluntarily cleaning the streets of discarded waste (Dalrymple, 2016). In 2016, a trend of picking up litter while jogging started in Sweden, giving rise to the term 'plogging'. By now, plogging's popularity has reached other countries, where it has, in the eyes of some, become a new fitness craze (Poole, 2018). This practice surely must appeal to a younger generation of health-conscious, exercising Chinese city dwellers, and its effects may be surprisingly positive over the long run.

Waste and O2O services

Based on the information I collected through interactions with Beijingers, it has become clear that the O2O companies have not been successful in their public relations work. None of my sources had ever heard of them, none of them had seen their advertising or seen them in practice. The sole exception were the RVMs operated by Incom, which many had encountered. Some actually used them, sometimes, and in general they liked the idea. Most damaging were the residents' observations that O2O services were not needed. They were satisfied with the solutions for the waste disposal problem that already existed and that worked just fine in their opinion. They did want to know more about safe ways to dispose of batteries, and of trustworthy services to help them dispose of used and/or unwanted clothing. A recurrent complaint was that despite their own desire to act and improve the environment, the residents did not know how. In their eyes, the government had failed to tell them what to do and had failed to find a solution for the waste.

4   The human factor – garbage pickers

In most urban areas, municipal departments are responsible for the disposal of solid waste. This waste is taken out of the city and transported to landfills or incinerator facilities. However, in reality, long before this garbage is disappeared completely, it is painstakingly – but not completely – sorted to remove anything that might be of value: metals, glass, plastics, paper, cardboard, textiles, aluminium, lead, steel, etc. The informal and unregulated practice of hand-sorting waste has a long history in urban China's pre-modern times and has continued into the present. It has become an important way to reduce the size of the waste stream and has provided the huge amounts of recycled resources and raw materials that industrial production needs. They have also reduced manufacturing and environmental costs of production considerably, as virgin resources do not need to be prospected, mined, and produced. Nowadays, the people who are sorting through the garbage of others belong to the groups of migrant workers who have moved from the countryside in waves since the 1980s to try to make a living and to become part of the city. They have effectively tried to escape the drudgery and lack of perspective of the countryside, to have more economic and life opportunities, to be recognized as full members of Chinese society and, at some stage, to even become registered as city dwellers – thereby gaining access to the many benefits that urbanites enjoy.

Migrant labourers

Who takes care of the waste of daily life in Beijing, apart from the workers employed by the municipal sanitation department? Who are these people sorting through other people's waste? The decollectivization of agriculture and the reduced control over the rural population in the 1980s – the result of a less strict implementation of the *hukou* legislation that tied people to their native areas – caused a vast pattern of migration to emerge, at one point making up more than 20 percent of the population (Wong, Li, and Song, 2007; Steuer et al., 2017). Some 100 to 200 million people are estimated to have left their rural homes since the 1980s (Solinger, 2006: 179). Decollectivization made superfluous labour power in the countryside both visible and problematic, as rural incomes became directly related to labour input. The large groups of underemployed labourers that had existed in the countryside could no longer find work, and families found the burden of supporting underperforming kin increasingly unbearable. During slow periods of farm work, peasants first made the move to urban areas, starting with the nearest townships and ending in the big cities and newly established industrial boomtowns along the Eastern Seaboard, such as Shenzhen. They found work in the urban building industry, which had an insatiable appetite for labour. Others were employed in the assemblage and piece-work production modes that thrived in the many new factories and workshops set up by foreign-owned, Hong Kong, and Taiwanese companies. After a short time, this migration pattern ceased to be seasonal and peasants started to move east and stay for good. The mechanism of chain migration enabled many people to move away from the land (Solinger, 1999). This itinerant labour power increasingly ended up in the metropolises, i.e., Beijing, Shanghai, Guangzhou, and others; in a later stage, second-tier cities such as the provincial capitals also became favoured destinations. Most labour migrants found employment doing casual or unschooled labour. This was not necessarily by choice, however: many cities barred migrants from a number of job categories (Béja et al., 1999a, 1999b; Zhang, 2002; Liu, Wong, and Liu, 2013).

This growth of rural-urban migration was linked closely to the spectacular demand for construction workers in urban areas where the scarcity of adequate living facilities increasingly felt like a burden (Guang, 2005). Agricultural labourers were seen as compliant, sturdy, and dependable workers, who were thus well-suited for the demands of construction work (Goldstein, 2006; Mobrand, 2006). Many of the waste pickers that were interviewed for this project started out as members of rural construction teams in the 1990s (Li, 2006; Interviews, 2017) before moving on to other occupations. They moved on either because cheaper labour power had replaced them, or because they calculated that waste picking offered better earning opportunities. By taking on work that most born-and-bred urbanites considered too dirty, too demeaning, and too poorly paid – even when these urbanites were themselves unemployed as result of lay-offs by their former employers, the state-owned enterprises (SOEs) – the migrant workers filled an essential niche in the urban configuration (Solinger, 2006; Goldstein, 2006). One can argue that the migrant labour force has functioned as a lubricant for China's growth.

Waste picking

The migrant worker-turned-scrap collector took over the work that was left when the formal recycling structure disappeared. In the process, they have developed collection patterns that are reminiscent of the routines followed by the 'pole carriers', 'big basket toters', 'small drum beaters', and

**Illustration 4.1   Collected PET bottles and assorted plastics waiting to be moved
to the next higher collection point for further processing, Haidian**

Author's photograph, April 2017

'large drum beaters' of the 1930s (Goldstein, 2006; Steuer et al., 2017). They
collect, bring together, or buy recyclable materials that they then transport
to transfer stations, where these materials are sold for a higher price; in
the process they earn a reward, or wages, for their labour input. Migrant
workers in transfer stations move the trash higher in the waste stream, until
it finally arrives at the recycling companies that turn it into resources. It
is hard and demanding work, consuming a lot of time, but even the small
mark-ups earned by the sorting and transporting of waste can generate a
total income that dwarfs the poor rewards that migrants still working in
the construction sector hope to receive. It is irrelevant to compare waste
picking with the income that can be received from rural work. Moreover,
as my sources stressed time and again, their independent status is a major
consideration for such migrants: one is free to do what one wants when one
wants to do it, and there is nobody who can change that.

I agree with the observations of Wu and Zhang (2016), who reject the idea
that these waste pickers are down and out, with no other job to turn to, and
almost by default gravitate towards waste picking (Hunwick, 2015). People
do not end up picking waste. Migrants do have agency and they are proud
of the work they do (Interviews, 2017). As a result, garbage picking looks less
like an earning opportunity of the last resort, and more like a way to get
rich relatively quickly and independently. This line of thinking is further
supported by the many stories making the rounds among migrant workers,

which tell of waste pickers who became the 'King of Glass', the 'King of Plastics', even the 'King of Steel and Aluminium', earning billions of yuan in the Beijing suburbs in the days when the numbers of waste sorters were smaller and there was more waste to pick (China Youth News, 2016; Zhao, 2016). The legendary exploits of these pioneers of waste sorting provide motivation and inspiration to the migrant workers (Illustration 4.1).

Changing one's source of income is often possible because native-place ties offered a chance to move elsewhere. In conjunction with occupational networks, native-place networks are particularly relevant in the waste-picking business. Encountering people from the same village, district, town, or province who are engaged in more lucrative work, for example waste picking, opens up new and more rewarding opportunities to earn a living (Mobrand, 2006; Wu and Zhang, 2016). This network of native-place ties has also resulted in serious competition and fights between networks concerning choice spots and types of work, leading to (informal and intergroup) agreements about which group engages with which type of waste (Zhao, 2016).

Some of these choice spots are the government units that still run their own recycling operations. These *danwei*, such as universities and other types of work units, classify and separate their garbage themselves. The moving and disposing of the recyclables is then contracted out to small operators, i.e., waste pickers, while the non-recyclable garbage is hauled off by the district or municipal sanitation department (Ensmenger, Goldstein, and Mack, 2005; Fieldnotes, 2017). I had the opportunity to observe the practice in one such unit, Peking University in Haidian District, in some detail. All day long, employees of the University's own sanitation department collect the garbage and waste produced by the estimated 50,000 people living and working within the unit (Interviews, 2017). Aside from the household and kitchen garbage of the departments, dormitories, dining halls, restaurants, and waste bins, the yield also includes the green waste (branches, leaves, duckweed, grass clippings, etc.) that is left from the process of maintaining the vast grounds and lakes of the university. All this waste is collected at a central point in the Southwest corner of the campus and sorted in a non-systematic way: some, but not all, of the plastics, cans, paper and cardboard, textiles, etc., are taken out and kept aside in the yard behind the collecting station. What is left is compressed in an underground container that is depleted, but not emptied, twice a day by a garbage truck operated by the Haidian Sanitation and Environment Bureau. These trucks deliver their loads to two incinerator facilities, one in Liulitun and one in Dagongcun, Sujiatuo, both located in Haidian District. One of the truck drivers emphatically said that Haidian waste needs to be incinerated in Haidian, suggesting

Illustration 4.2   Paper and cardboard at a Xiyuan Bridge collection point, Haidian, waiting to be moved for further processing

Author's photograph, 28 March 2017

that the district is responsible for its ultimate disposal (Interview, 2017). The recyclables are carted off by (informal) tricycle operators. But aside from this seemingly organized system of waste disposal, with its deals and agreements that suggest a formal division of labour, there are other, smaller and independent operators at work as well. Some of them even have access to some sort of warehouses on the campus, keeping their loads in the often-disused air-raid shelters under the buildings. They carefully pre-sort the waste before they cart the recyclables away to the collecting station. I have not been able to establish how these more informal waste pickers are able to enter the closely guarded and patrolled university grounds, or how the contacts and contracts with the less informal ones are established or structured (Fieldnotes, 2017) (Illustration 4.2).

Native-place ties

Given the informality of the waste-picking sector, it is hard to estimate how many people are engaged in it in urban China. Some estimates mention the number five million, forming part of the 20 million people that are employed in one way or the other in the recycling industry nationwide (McKinsey & Company and Ocean Conservancy, 2015; Inverardi-Ferri, 2017). But these estimates are hindered by the same problems that are encountered when using statistical data, as discussed in Chapter 1 (Linzner and Salhofer, 2014). Comparing various studies, Linzner and Salhofer have calculated that the number of people sorting waste in Beijing ranges anywhere between 130,000-300,000 (2014: 903). A more recent estimate, from 2016, suggests that some 100,000 of them were left in 2014 (China Youth News, 2016). The influence and strength of the native-place ties are visible in the type of work they engage in and the locations they are active. The people who collect waste from Beijing's streets are predominantly from Sichuan Province. In 2016, there were at least 40,000 Sichuanese from the city of Bazhong (Dazhong District) alone working in Beijing. Most of the informal waste collectors I talked to were from Bazhong. The majority of the collection depots in Beijing that are still in operation are run by people hailing from a single district in Henan Province, i.e., Gushi district (China Youth News, 2016; Zhao, 2016; Tong and Tao, 2017; Interviews, 2017).[12] In 2016, some 17,000 depot bosses from Henan were active in Beijing; the ones I encountered and spoke with were all proud natives of Gushi (Zhao, 2016; Interviews, 2017). People from Hebei Province

12   https://zh.wikipedia.org/wiki/%E5%9B%BA%E5%A7%8B%E5%8E%BF

have taken control of waste sorting and collecting in the area outside of the Fourth Ring Road. And migrants from Jiangsu Province have a monopoly over the illegal recycling of cooking oil, also called gutter oil (Zhao, 2016).

Henan is considered one of China's most densely populated provinces. Over the centuries, large numbers of migrants have left it, not only within China but moving out into the wider world as well. Gushi district is in the area that was severely affected by the famine of 1958-1960 that followed the Great Leap Forward campaign, which lead to more than 100,000 casualties in Gushi alone. Formerly an agricultural base, its labour force became redundant to a large extent after the Reform policies struck root in the early 1980s and decollectivization took place. This caused an exodus of labourers to provinces and cities such as Jiangsu, Guangzhou, Shanghai, and Beijing looking for non-agricultural jobs (Li, 2006). As with other former agricultural production areas that witnessed large numbers of their labour force move to urban areas, the remittances of these members of the floating population have contributed extensively to the present relative affluence of the district.

The structure of the waste stream

Roughly speaking, there are six layers of recyclers in Beijing, each successive layer adding value to the recyclable waste. These layers show a similarity with the make-up of the informal recycle systems that are in place elsewhere in the world (Ezeah, Fazakerley, and Roberts, 2013: 2512-2513) and consist of:

1   the scavengers, who freely pick up recyclables on the streets, in shopping malls, subway stations, residential communities, and landfills, dumps, or transfer stations. Alternatively, the free-moving scavenger is a waste picker who has come to an agreement with a residential community, restaurant, or other type of business to have exclusive access to the recyclable waste generated there. These agreements are often informal and can be quite costly, depending on the richness of the pickings. They can also be controlled by crime gangs that exchange access to junk for payment (Ensmenger, Goldstein, and Mack, 2005: 126; China Youth News, 2016). Other types of collaboration emerge between the drivers of municipal garbage disposal trucks and pickers, such as sharing valuable recyclables (Williams, 2014: 216; Kao and Lin, 2018).

2   the itinerant waste buyers, who collect recyclables door-to-door, paying for the items they pick up. They sell the waste to the medium or large redemption depots. Many of them use tricycles or bikes to transport what they have collected (Norcliffe, 2011) (Illustration 4.3).

3   the small community waste-buying depots, operated by people who
    have paid a land usage fee to the community management committee in
    whose jurisdiction they operate. Through this, they have a permit from
    the local government to recycle. The small community waste-buying
    depots occupy some 5-10 square meters and do the waste collection in a
    community. These depots often resemble the portacabin structures that
    are used in the Hong Chao pilot project discussed in Chapter 3. Their
    operators stay in place and wait for people to bring them recyclables.
    In addition, they occasionally do some door-to-door recycling when
    people call them. Aside from these in-community depots, tricycle or
    truck-operating junk buyers are active curb-side, on street corners, and
    in small plots tucked away between the residential communities.

4   A variation on the previous layer is that of the recycling stand. These
    stands can be either of a periodic nature and independently run, or
    more regulated and contracted by the municipal government. Both
    operate on the sidewalks outside of communities. The periodic stands
    transport their materials to recycling markets or directly to the recycling
    plant, while the regulated stands sell directly to the recycling plants
    (Ensmenger, Goldstein, and Mack, 2005; Fieldnotes, 2015, 2017).

5   the medium/large redemption depots and recycling markets are where
    the collected recyclables from the community depots are brought to-
    gether; the larger the depots, the further removed they are from the city
    centre. Over time, these depots have moved from the area between the
    Second and Third Ring Roads to between Fourth and Fifth Ring Roads
    to beyond the Fifth Ring Road. The operators of the trucks transporting
    the separated recyclables to ever-larger depots can be subjected to
    harassment and extortion from crime gangs (Ensmenger, Goldstein,
    and Mack, 2005: 126; Johnson, 2013a; China Youth News, 2016).

6   the recycling companies or factories are the final destination, where the
    waste ultimately is turned into new raw materials that can be reused.
    Plastics are processed in Wen'an, south of the city; Bao'an takes in the
    metal; paper ends up in Baoding; and glass is sent to Handan. Ideally,
    these companies and factories operate with pollution control equipment
    under the supervision of local governments, but this is not always the
    case (Ensmenger, Goldstein, and Mack, 2005: 121-124; Wilson, Velis, and
    Cheeseman, 2006; Linzner and Salhofer, 2014; Zhang and Wen, 2014: 993-
    994; China Youth News, 2016; Steuer et al., 2017; Collective Responsibility,
    2017). Under newly promulgated regulations to improve the air quality of
    Beijing, recycling factories in Handan, Baoding, and elsewhere have been
    ordered to cease operations, some only temporarily, some completely.

Illustration 4.3   A bicycle loaded with collected junk, Haidian District

Author's photograph, 14 March 2017

The competition between informal waste pickers, augmented by a growing number of newcomers that continue to escape from the rural areas, and the expanding groups of original urban inhabitants who need scavenging to add to their income, is growing increasingly bitter. This latter group includes the old-age pensioners who can often been seen scavenging in an attempt to increase their pensions, particularly in parts of the city that are considered less well off (Liu and Wu, 2006; Goh, 2009; Griffiths, 2014; McKinsey & Company and Ocean Conservancy, 2015). The frequent conflicts between informal waste collectors and the elderly can particularly be witnessed near subway stations, where many prospective metro users want to get rid of their (almost empty) PET bottles in anticipation of the security checks conducted before boarding. Other potential battlefields are crowded spots that promise lucrative hauls of recyclables. These destinations that draw large numbers of domestic and foreign visitors all year include the Happy Valley Water Park (Huanlegu), the Beijing Zoo, Yuanmingyuan and the Summer Palace (Yiheyuan), the Forbidden City, the various Olympic structures (Bird's Nest, Water Cube), etc. Less obvious destinations such as parks and even the neighbourhoods where the campuses of Peking and Tsinghua Universities are located can turn into potential places of contestation between waste collectors due to their large numbers of visitors (Fieldnotes 2015, 2017).

The *Suzhi* of waste pickers

These out-of-town waste pickers are not looked upon kindly by either Beijing officials or residents. They engage in low-level work that city dwellers refuse to do and are considered to lack *suzhi* ('quality, culture'). This concept is discussed in more detail in Chapter 5. Waste pickers and their work reflect badly on the ideas of modernity that the government entertains (Zhang, 2002; Inverardi-Ferri, 2017; Interviews, 2017). There is a growing resistance to informal trash collection in neighbourhoods where incomes and living standards are high. The higher the educational levels of the residents, and the higher the status potential of the residential community, the more outspoken the prejudice against migrants (Tse, 2016; Collective Responsibility, 2017). Waste workers have a bad reputation and are immediately associated with criminal behaviour by the 'fine' townspeople (Xu, 2008). They mess up the grounds of the community with their incessant digging through waste bins; they cheat when they buy valuable junk; and they cannot be trusted. There have been scores of reported cases of, as well as rumours about, migrant waste pickers who, being unable to find any valuable waste, have turned to stealing manhole covers, steel fences, transformers, and steel and copper cables – but that should not disqualify the whole contingent per se (Zhao, 2016). Chinese crime statistics report that a majority of police arrests concern migrants, and yet migrant workers are also more likely than residents to be the victims of crime (Sun et al., 2013). Aside from not being accepted in the urban environment for these reasons, administrative discrimination at all levels and popular distrust and suspicion have made it almost impossible to build a life in urban areas for those who were not born there. Having an urban *hukou* provides a passport to living in the city.

The Household Registration System regulations (*hukou*) that were introduced in 1958 basically divided the population into a very large rural and much smaller urban section, with the intention of keeping the rural inhabitants from moving to the cities. The main reason for the introduction of these rules was the nationwide famine that was threatening at the time. The rules were intended to protect the cities from being overwhelmed by starving peasants and to safeguard the meagre supplies of food that were available for city residents. These regulations brought to an end the freedom of internal migration and residence that had originally been written into the State Constitution that was in force at the time. The *hukou* system has seen some reforms and changes in the recent past, of which one, less strict enforcement, enables rural workers to move to the city temporarily. However, the restricting and restrictive rules built into the system basically continue to exclude migrants from full citizenship (Chan, 2009). For those who moved to urban areas after the reforms started, it remains extremely troublesome to find housing, rent a shop location, register vehicles, obtain a local cell phone number, register a business, get a loan, etc. The *hukou* system continues to deny migrant workers a number of basic entitlements that urban residents can lay claim to, thus making the managing of family life, i.e., renting housing and sending children to school, equally cumbersome. Even though further reform of the *hukou* system has been in the works for years, little progress has been made. Despite their essential role in the city, urban residents and officials continue to look down on migrants, seeing them as population pollution (Goldstein, 2006; Zhang and Li, 2016; Goldstein, 2017: 178).

Urban villages

As a result, migrant workers have been forced to converge on the outskirts of the city, to the hybrid areas where town and country merge, a space also known by the term *desakota* (Béja et al., 1999a: 29; Davis, 2006; Yates, 2011). Native-place ties and chain migration processes have resulted in the concentration of people moving from the same districts, towns, and provinces in the same settlements, hence the appearance of migrant enclaves within Beijing. In 2008, 876 of these were identified but not included in the urban statistics (Zheng et al., 2009: 428). Similar migrant enclaves emerged around other tier-one, -two and -three urban areas as well. The more famous ones in Beijing – famous due to the broad interest that they attracted over the years from both Chinese and Western researchers – were known as 'Henan Village', formerly located in a part of Haidian District that still was seen as a suburban area in the 1990s, and 'Zhejiang Village' in Fengtai District, which housed some 100,000 migrant entrepreneurs from Wenzhou and hired-wage workers from other regions. Some other enclaves gained fame for a while as well. What distinguishes these enclaves is that they were all named after the location that the majority of their inhabitants hailed from (Béja et al., 1999a; Zhang, 2002: 317). These configurations, villages-in-cities or 'urban villages', also retained strong village features (Zheng et al., 2009; Shin and Li, 2013; Zhao and Zhang, forthcoming). The migrant centres evolved into small worlds unto their own: they had and have their own supermarkets, eateries, laundry shops, vendors, and other stores, some run by self-employed migrants, that supply services and cater to tastes that make their inhabitants feel like they are at home though abroad (Xu, 2008; Zheng et al., 2009; Loyalka, 2012). Each of the enclaves rendered different services to Beijing's economy and society: Henan Village provided labour and served as a market for garbage collection, while the people living in Zhejiang village mainly worked in the garment industry. At one point, Zhejiang village even acquired a monopoly in Beijing and North-eastern China over the clothing produced there, which was marketed as original Zhejiang Province items (Béja et al., 1999a; Zhang, 2002: 317; Ou, 2011).

Pushed out of the city and finding it nearly impossible to find affordable housing, migrant workers were often forced to rent accommodations from local peasants who used to work the soil. In the period 2005-2017, I have seen the area known as Sijiqing, located south of the Summer Palace and between the West North Fourth and Fifth Ring Roads, evolve in this process (Fieldnotes 2005, 2007, 2008, 2010, 2012, 2015, 2017; Zheng et al., 2009). Sijiqing used to be an agricultural base of production, with rural villages, farmland, orchards and tree nurseries, and only a few small migrant communities. Over the years, habitation increased and residential communities for Beijingers emerged; car dealerships sprung up; both shady and more respectable karaoke bars set up shop; informal night markets appeared, attracting thousands; and the number and size of the migrant communities congregated there grew. Over time, the area slowly but surely gentrified, connecting it with the network of public transport, forcing out the establishments and inhabitants that were considered less savoury and replacing them with more appropriate people and businesses, including hotels and golf courses. By 2017, the area has become a fully integrated part of the larger district, particularly the area bordering both sides of the Fourth Ring Road. The migrant population that initially found ways to establish itself there has been forced further west and north beyond the Fifth Ring Road, learning about alternative 'cheap rental through word of mouth, via friends or relatives, or by reading casual advertisements in the village' (Ou, 2011; Liu, Wong, and Liu, 2012: 1233).

While most agricultural workers in suburban areas lost their rights to toil the land in the process of Beijing's outward movement of urbanization, they discovered a way to earn some (extra) income by renting out rooms in structures built on the parcels of land reserved for housing to migrants at affordable prices (Liu, Wong, and Liu, 2012; Zheng et al., 2009: 426). Many came up with strategies to expand the rental possibilities of their homes, such as building floors on top of existing dwellings, but this seriously decreased the safety of those renting these rooms. Others built housing compounds (*dazayuan* 'mixed buildings') (Xu, 2008; Zheng et al., 2009; Ou, 2011; Shin and Li, 2013). Yet even communities such as these were demolished in the

end. The famous Zhejiang Village, for example, was finally levelled in 2006
to make room for construction work related to the 2008 Beijing Olympic
Games (Xu, 2008: 648; Shin and Li, 2013); the people working there had, in
the meantime, spread out to almost every city in China (Zhang, 2002: 317).
The last inhabitants of Henan Village were still working in the recycling
trade in 2017, in a few tiny corners that were left of the original enclave. They
too had been given notice that they would have to move elsewhere before
the end of the year, due to the construction of the residential quarters for
Tsinghua University staff that was already encroaching on their work space
(Fieldnotes, 2017; Interviews, 2017).

Hongfu Yuan

The destruction and ultimate disappearance of these migrant communities is
not predestined. Zhao and Zhang (forthcoming) have traced the development
of the northern Beijing village of Zhenggezhuang, located between the Fifth
and Sixth Ring Roads. They convincingly argue how informal and even
partly illegal communities such as these can evolve into 'respectable' gated
communities, as in the case of Hongfu Yuan, which currently accommodates
migrant residents, college students (from the branch campuses of the Beijing
University of Posts and Communication (BUPC) and the Central Academy
of Drama (CAD)) and 2456 indigenous villagers. The gated community is
the result of the transfer and pooling of the land-use rights of the original
villagers into the Hongfu Group in 1999. This group is responsible for the
governance of the community and administering its further development
and diversification, which includes the construction of a tourist resort
(Wendu Water City) with hotels, a golf course, and a shopping mall. Hongfu
Group has been able to entice the Beijing Bus Company to provide bus
services to the community, including one rapid transit bus line, linking it
with the city proper.

Lajicun ('Junk villages')

Zhenggezhuang is an obvious success story for the informal communities
that are emerging around the city, and about the ways and strategies for
those working in the informal economy to carve out a space (Liu, Wong, and
Liu, 2012). It is not representative of the many other suburban population
centres where economic processes have forced the scrap collectors and

their recycling stations over the years. They have been pushed out by the same housing estates and shopping arcades where the recyclables that they collect, separate, and sell are produced in ever-increasing amounts. The scrap collectors and recyclers have taken refuge in 'junk villages', which take their name from the junk and recyclables that are concentrated there and the scrap markets that they organize. The 'junk villages' often operate the legal, semi-legal, and illegal garbage dumps in their vicinity as enterprises (Kao and Lin, 2018). These villages surround the urban area of Beijing like a virtual Ring Road (Wang, 2011). The process of forcing the activity of recycling ever further out of the built-up areas has made the process of carting away garbage increasingly time- and energy consuming, irrational, and expensive.

While dealing with domestically produced garbage has thus been made ever more difficult, the situation has been compounded by another factor that imperils the functioning of these 'junk villages'. It also serves as an illustration of the way in which informal Beijing migrant workers have been made part of the global nature of the junk, scrap, and recycling enterprise (Medina, 2011). Located as they are near the garbage dumps that they operate as enterprises, for years the villagers have been processing the garbage of the world that China has both produced itself and imported from elsewhere (Goldstein, 2006). According to government figures, China imported 49.6 million metrics ton of waste in 2015 (Agence France Press, 2018). As Adam Minter (2013a: 85-88, 2013b, 2015) has explained, for an extended period of time the shipping containers that carried export products from China to consumers worldwide also brought back the garbage produced by those consumers, for a reduced rate. The shipping companies could not allow containers to return empty to China. The export of finished products and the import of junk created an almost perfect loop. The extraction of much-needed raw materials and recyclables from the waste produced elsewhere has thus allowed the Chinese industrial export machine to attain the level it has today, churning out more and more products that are consumed globally, and in the process producing more waste that has no place to go. By importing and using the refuse of the world, China has been able to preserve large amounts of the nation's virgin resources, energy, etc. In a perverse way, this process has therefore contributed to an improvement of the environment. In the summer of 2017, however, the CCP Central Committee and State Council adopted the '2018-2020 Action Plan for Full Implementation of the "Implementation Plan for Banning Foreign Wastes and Advancing the Institutional Reforms on the Management over Importation of Solid Wastes"', making it clear that it was no longer interested in recycling other nations' garbage. The ban on the import of foreign waste, which took effect on

1 January 2018, is to be fully implemented by 2020 (State Council, 2017; MEE, 2018). This decision was not only based on the fact that the country produced enough garbage of its own; it was also seen as promoting the development of the circular economy. The Plan, moreover, gives opportunities to the governments at various levels to further regulate, and make healthier, the domestic recycling sector, phase out informal waste pickers, etc. The ban has caused a huge headache for the international garbage-producing and -recycling industry (Agence France Press, 2018).

Closing down the waste villages

A number of the waste villages surrounding Beijing have become important sites for research by Western academics. Whereas in the 1990s these researchers studied the emerging entrepreneurship amongst the migrant population in Henan Village and Zhejiang Village, in the first two decades of the millennium their attention has shifted elsewhere, focusing on marginalized people leading precarious existences by sorting through waste. Many of these villages remain nameless, but the villages of Dongxiaokou in Changping District (Zhao, 2016; Steuer et al., 2017; Goldstein, 2017; Inverardi-Ferri, 2017), Huangcunzhen in Daxing District (Steuer et al., 2017), Lengshui (Wu and Zhang, 2016) and Picun ('Leather Village') in Chaoyang District (Huang, 2015), and others have become household names in the field. Although the activities of the villagers and their living conditions have been used by municipal departments to further their own agendas, such as the promotion of incineration, they continue to be seen as unwanted side effects of modernization that the city wants to get rid of. Liu, Wong, and Liu, for example, refer to the 2010 city-wide adoption of new residency control measures by Beijing Municipality as embodied in the 'Shunyi Model' and the 'Daxing Model' (2013: 366-369). Starting in 2000, these two districts focused on building higher-end industries that require a higher skilled workforce, who in turn demand higher-end housing. This approach has slowed down the increase of low-wage migrants moving into the districts and has gentrified the areas, as illustrated by the example of Zhenggezhuang.

In late 2017, after a fire broke out in one of the many illegal structures in the Xinjiancun section of Dongxiaokou, the Beijing municipal government and other official departments started a number of concerted campaigns to drive out migrant workers, now termed the 'low-end population', and force them to return to the places they hail from. This can be seen as the final act in a long series of attempts to regulate these and other waste villages (Chen, 2014; Battaglia, 2017). According to Michelle Yates, this is the next stage of waste management and disposal, which 'recently has taken the form of "eviscerating urbanization", i.e., the cleaning up and removal (management and disposal) of slums and slum residents' (2011: 1687). What does this mean for the garbage collection, retrieval, and recycling sector? Or, to put it differently, who will take care of urban garbage now?

Types of O2O employments

This is the point where the activities of the O2O companies introduced in Chapter 2 come in. They are well-organized, boast of a well-behaved workforce in designated uniforms, and promise secondary pollution-free waste disposal. In short, they are just the types of waste collectors that the government would like to have. But is this work attractive or available to the informal waste pickers? What does it mean for them to become an employee of such a company, someone hired to pick up the recyclables that have been reported by citizens using dedicated apps on their smartphones? And what does it take to become such an employee? My attempts to have such questions answered by the O2O companies themselves did not lead to any results. Instead, I engaged in extended research on the internet pages of three of the companies operating in Beijing that listed vacancies. This has resulted in the following information concerning the types of employees they were looking for at the time. It also sheds light on the diversity and breadth of the activities that O2O companies are engaged in or planning to branch out into.

Taoqibao

Taoqibao only has openings for 'recyclers'. The Taoqibao website[13] defines what it expects of 'recyclers', as it calls its potential employees. They must be aged between 18-35 years, and they should have completed high school, secondary school, vocational high, technical or higher, but college education is preferred. Their facial features should be correct: they should have a good image, a smooth expression, and be clear thinking. They should be able to work hard, have a strong sense of professionalism and responsibility, and be interested in the company's long-term development. The website does not explain what working as a 'recycler' means. It is not clear what

---

13   http://web.taoqbao.net/contact_us.html. The page is no longer active.

showing interest in the long-term development of the company means for the individual applicant. But the requirements show that migrant workers are certainly not the most likely group of people to be hired for this position.

In return, the employee earns a fixed monthly salary of 2200 yuan; this can increase through commissioned (piece-rate) wages to an average income of 4000-7000 yuan per month. The company offers social insurance covering a retirement pension, medical insurance, unemployment insurance, work injury insurance, and maternity insurance. It also provides employees with birthday presents and a variety of holiday gifts. Work clothes are also provided free of charge throughout the year, but the website does not indicate what these include. The company organizes a variety of staff activities, ostensibly to create a true company spirit. Aside from full-time employees, the company is also interested in contracting freelancers, people who want to make money in their spare time. According to the website, the earning opportunities are almost similar to those of white-collar workers.

Zai Shenghuo

New Living (Zai Shenghuo) is looking for the largest number of employees, which is in line with their plans to expand their activities to other fields, as discussed earlier. Their recruitment page[14] lists the following vacancies. All job descriptions include the provision that wages are paid on the 10th of each month, and that no money will be withheld. The need to specifically spell out this provision indicates that some companies are slow in paying, or are holding back part of the wages for whatever reason.

- App promoters, i.e., persons who urge others to download and use the app. For this job, people between 18 and 28 of both genders can apply; the household registration status of the applicants is not relevant. There is no fixed salary mentioned, although the income range is quoted at 4000-12,000 yuan per month, based on piece work rates. There is no indication how these rates are calculated. The job is on the basis of a contract and comes with insurance. The company promises 'career development space'.
- Couriers (tricycle operators), who are responsible for daily delivery services and maintenance jobs. They should be 18-35 years old; household registration is irrelevant; and education level should be at least junior high school. Wages are based on a monthly salary (level not specified), augmented with a performance salary, competitive bonuses, and other components. The average income can reach 5000-8000 yuan per month but is dependent on

14 http://www.anewlives.com/jionUso1.html

piece-work rates. There is no cap on income. Pre-job training is provided by the company, which also offers staff career-development space. Work is done on the basis of a formal labour contract with insurance. Free work clothing is provided, and implements/tools used on the job are provided free of charge and deposit. The company frequently organizes recruitment drives for promotion, training, and internships.

- Mobile phone detection engineers (home), who test mobile phones, tablets, notebooks, cameras, and other electronic equipment. The company is looking for people with a college degree or higher, under 30 years of age; they should be handsome, healthy, and hard-working, have a strong sense of responsibility, a good team spirit, master standard Mandarin Chinese, and have good communication skills. Pre-job training is provided by the company, so previous knowledge of the technologies is not required. The company offers a staff career-development space. Work is done on the basis of a formal labour contract and includes insurance. No salary specification is given. Note that the requirement to communicate in standard Chinese has the potential of barring migrant workers from these jobs.
- Air conditioning replacers. They should be 20-35 years old and have at least one year of experience working in the air conditioning business. The monthly salary is based on a basic salary plus commission, totalling 4500 yuan per month, dependent on piece-work rates. The company offers staff career-development space. Work is done on the basis of a formal labour contract and includes insurance.
- Sorters, who sort garbage at collection stations. The company looks for persons 18-40 years old, hard-working, with a strong sense of team spirit, and expects long-term commitment. The monthly salary is 4000-6000 yuan, but this is without a contract or insurance.
- Truck drivers, who should be 25-40 years old, hold a B2 driver's license (freight certificate) or higher, and have more than two years of actual driving experience. They must be familiar with the local road conditions, have a good driving record without major accidents and traffic violations, and have a strong sense of security. The salary amounts to 4000-8000 yuan per month, on the basis of a formal labour contract including insurance.
- Warehouse managers should have an education record of secondary school or higher and be aged 20-35 years; they should have more than one year of relevant work experience. The salary is 4000-6000 yuan per month on the basis of a formal labour contract including insurance.

On top of that, the company also provides free staff quarters on the premises, with a kitchen and independent bathroom.

Incom

Incom advertises only a few job openings on its website,[15] but none concerning the collection of recyclables. They are more like back-office occupations that deal with tasks related to running the company itself. This is remarkable, as Incom was the first company to boast about hiring informal waste pickers to stop the competition over recyclables. They have declared that they have 300 garbage collectors under their employment, but it remains unclear how they were recruited and hired. The job openings they do offer on the website range from technical service engineer, Image Recognition Senior R & D engineer, and Senior Electronic R & D engineer, to environmental Research Fellow. All these functions typically demand higher education degrees, ranging from BA to MSc levels; for none of these employment opportunities is an indication of the salary range given.

After comparing the vacancies that the three O2O companies advertised, one must conclude that the informal waste collectors, who usually hail from the cohorts of migrant workers, do not seem to fit the demands. The recruiters are looking for human resources that have an urban registration, finished school education at the secondary, vocational, or higher level, speak fluent standard Chinese, look pleasant, and so on.

The appeal of working for an O2O company

The O2O companies have not changed the status of the waste pickers. The latter have not become stakeholders in the process of recycling, a development that some consider a necessary step for their acceptance and inclusion (Tong and Tao, 2016; Goldstein, 2017). Chen Liwen, formerly attached to the Green Beagle environmental NGO, has vividly recorded the endless toils of Mr. and Mrs. Ma in her 2012 video 'A Beijing Recycler's Life' (Chen, 2012). Would the Mas consider joining an O2O outfit? Would they see it as an option to gain social capital (Prasad et al., 2012)? Would it change their personal connections with residents, which have been witnessed and reported on so often (Prasad et al., 2012; Minter, 2013a;

---

15   http://www.incomrecycle.com/recruit/

Wu and Zhang, 2016)? Are the garbage pickers employed by the O2O companies competing with the 'freelancers' who are still very much active? Have there been any 'waste wars' between pickers from different O2O companies? And where do the old-age pensioners who pick garbage fit in this scheme? Can they join an O2O company? Are they willing to do so, and are there examples, other than the ones suggested by Chinese media reports (Zheng J., 2017)?

When interviewing waste pickers, collection station bosses and workers, and regular sanitation workers about these topics, they were initially amused by my questions and then invariably scoffed at my suggestion that they join an O2O company. Was their lack of an urban registration, education, and professional ability not the reason why they ended up in the recycling trade in the first place? – they argued. And did they, who had acquired their knowledge in practice, really need to be trained to do a job that required no training? – they continued. And why would they need to raise their level of providing service to their customers? None of them considered working for an O2O company to be a serious option. They did not recollect encountering any O2O workers and thus they were not considered competitors. They also saw the work and income potential of working for an O2O company as unfavourable, mentioning various reasons. First of all, both the pickers and the station bosses and workers considered the pay much too low, uttering lines like 'If you are willing to work hard, the [independent] income is much higher', 'This is only for fancy kids who do not know what hard work is', and 'They are only interested in high-end waste'. Only Mr. Li, a 53-year-old waste picker from Bazhong, seemed to like the idea of working for a company, but he assumed that it meant earning less and therefore decided that it was not worth his while. While agreeing that the informality of their work could mean that they earned less on some days, most pickers were convinced that they could do much better than the 2000 yuan-a-month that most of the lower-level O2O jobs offered. Indeed, the correlation between hard work and high income was mentioned most frequently, and with considerable pride. Some bosses let on that they had been approached by O2O companies but had turned down their offer of a more formal type of cooperation. The bosses said that the 3000-4000 yuan the companies had offered them did not compare to the amount of money they could earn themselves in a day by depending on their own labour power. Mr. Wang, a 40-year-old boss from Gushi, Henan, told me he sometimes bought some of the junk that the O2O companies offered him, but he only did so when he could get it for a good price. In other words, O2O companies were just business relations

Illustration 4.4   Discarded bicycles awaiting transport at a Haidian District scrap collection point

Author's photograph, 14 March 2017

like any other, nothing special. He did not see his dealings with them as a formal type of employment, but just ordinary business (Interviews, 2017) (Illustration 4.4).

Neither pickers nor bosses had heard about the smartphone apps that the O2O companies had developed, and they wondered what they were. When I showed them how some of them, like Bangdaojia and Taoqibao, worked, they were not impressed. The pickers did not like the idea that they could be summoned to a residence to pick up waste. This actually contradicted the stories reported in a number of Chinese media articles that referred to the convenience the apps provided for the pickers (Zhang, 2016; Zheng J., 2017). A few of the bosses mentioned that sometimes long-term clients with which they had built profitable relations, like companies or restaurants, would contact them in the old-fashioned way, giving them a phone call when they had junk for sale. They usually decided on the spot, depending on the price they had to pay for such offers, and could not be bothered by the idea of set prices. By and large, my sources sounded suspicious of the apps. They were convinced that these were just tricks engineered by the O2O companies to get government subsidies; using apps would not change anything in the business of collecting and selling waste (Interviews, 2017).

Then there was also the issue of the informality of the work they currently did. As individual workers, they had more freedom and rights to decide the prices they bought and sold for, the workload and time they wanted to invest – even though the O2O job descriptions did suggest piece rate payment (Interviews, 2017). None of my sources indicated interest in joining something like an actual company. 'My income solely depends upon myself. If I work today, I get paid', was the type of answer they regularly gave. The possibility that joining an O2O company would provide their work with some protection from external pressures, like governments or the feared and hated city management officers (*chengguan*), did not occur to them. In fact, the waste pickers did not mention feeling any pressure at all from these corners. Some of the bosses said that these types of pressure were precisely the sort of grief they had to deal with, but it all came with the territory. The trouble of finding new vacant lots for their separation and collection activities gave them the biggest headache. They were under constant pressure to move further out of the city, as their collection points were taken over time and again. The prospect of wearing a company uniform did not appeal at all and even seemed to embarrass my sources a little, although they were partial to the idea of getting some sort of work clothes for free. They were not at all attracted by the promise of social insurance that some of the O2O contracts offered, as they were convinced that the premiums would be taken out of the wages. The only thing they liked about the O2O employment, particularly in the opinion of the waste pickers, was the chance to get free tools, like a cart. In 2001, a tricycle cost 200 yuan to use; by 2015, that had gone up to 2000 yuan, and in 2017, the costs had gone up again (Goldstein, 2006; Hunwick, 2015; Interviews, 2017).

As for gaining status and being more accepted by the native Beijingers by becoming a regular employee, they were not convinced that joining an O2O company was the right way to accomplish that. After all, they 'had no culture', they lacked *suzhi*. They wondered whether they actually wanted to do all of this in return for gaining acceptance; rather, they indicated that the respect they had or could gain in their own communities back home mattered more to them than acceptance in Beijing. All of them were positive that joining an O2O company would help them cut back on the expenses of acquiring the various permits that were required for their work, but they were also certain that going the O2O way would not help them get access to an urban registration (Interviews, 2017).

The attitudes of the regular sanitation workers differed from those of the informal waste workers in many respects. Most of the regular workers were born and bred Beijingers, so the prospect of joining an O2O-company to get an urban *hukou* had no appeal for them whatsoever. Moreover, they already worked for the sanitation departments of the municipal or district governments, or for residential committees. Although they were engaged in lowly work, it still afforded them face and respect, as they held an official position of a sort. All of the sanitation workers I interviewed were proud of their work and their contributions to society. They wore their uniforms with pride, as they distinguished them from others and gave them a formal position in the wider scheme of urban society. As they drove their garbage trucks or tricycles to collect junk, all provided by their municipal employers, the idea of having free tools of the trade at their disposal had no appeal at all (Interviews, 2017).

Informal versus O2O waste picking

One can say that the O2O companies' claims that employing waste pickers had a positive effect on the amount of junk they were able to gather seem dubious. Not one single O2O-employed waste picker, and only one company-designated vehicle, was encountered during my three months of research. Factoring in the fact that the O2O companies' base of activities was in other parts of the city, it still came as a surprise that the informal waste pickers and bosses were mostly unfamiliar with their existence and work. Although an informal information network among those employed in waste picking exists, it seems that O2O companies and their operations did not merit a mention there.

Likewise, working for an O2O company clearly held no appeal for the pool of labour power that the O2Os would like to tap. Waste pickers and bosses consciously favoured the informal aspects of their work, the ability to earn more if one worked hard, and not having to tolerate some supervisor looking on. From the perspective of the people working the waste, O2O companies were irrelevant.

5   Educating the people

Measures to reduce the amount of waste produced and to organize the collection (and recycling) of garbage are only parts of the solution. People need to be made aware of the problems involved in waste management. They need to appreciate that they must begin to produce less garbage, and they need to learn about less harmful ways to dispose of generated waste. This calls for continuous educational efforts by various actors. These educational activities were part of my research project in spring 2017.

Some of the topics discussed with respondents involved in educational work were the following: Which educational projects have they undertaken in the past, and which have they developed recently? Where did the projects take place, and what results did they have? Which means and media have proven to be the most effective in bringing about changes in behaviours and attitudes, i.e., printed media, public service announcements, special television programming, etc.? Are small-scale local events more effective than larger ones? Does President Xi Jinping's Chinese Dream campaign (中国梦), which started in 2012, have anything to say on waste, individual behaviour related to waste, recycling, etc.? As the Dream campaign seems to be quietly receding into the background, has anything else taken its place where it concerns garbage and its disposal? Finally, how effective are the regular educational posts that appear on the social media sites run by the recycling and O2O companies themselves? Do they contribute to changing the behaviour of consumers or subscribers? Who develops, designs, and produces these postings?

Suzhi

Before turning to these questions, it is necessary to devote attention to the topic of *suzhi*, or 'human quality', 'culture', or 'being cultured', that has been referred to frequently in the preceding chapters. Concerns about the (perceived) quality of oneself and others permeate evaluations of people's behaviours. Invoking *suzhi* has discursive power and confers authority on the speaker, no matter whether s/he is a government official, an ordinary urban residential community member, or a migrant labourer. It sets the speaker apart from the person spoken about, usually in a positive sense. While people indicate that others do not have (*meiyou*) *suzhi*, those thus identified as lacking an essential quality usually attribute this to *meiyou wenhua* ('not having culture'). People can have bad *suzhi* or good *suzhi*, indicating behaviours that an individual in society considers right or wrong. Having high or low *suzhi* signifies behaviour that is influenced by one's class background, education, and training (Upton-McLaughlin, 2014). The use of *suzhi* and its implications are reminiscent of the importance that class status or *chengfen* (family origin) had for determining a person's employment prospects, social status, and prestige in pre-Reform China (Li, 2010: 146).

In the past ten to fifteen years, '*suzhi* has become increasingly central to dynamics of culture and governance in China' (Kipnis, 2011: 290; Sigley, 2009). Though not a new term, concept, or goal to aspire to, the explicit concern with *suzhi* made its first appearance in the 1980s, when China prepared for the drive to modernize and open up to the outside world. It coincided with the introduction of the One Child Policy (OCP), where the harnessing of population growth was seen as part of a process of planning, social engineering, and eugenics that would end China's backwardness. Having a smaller population would make its people strong and would turn quantity into quality (Sigley, 2009; Anagnost, 2004). It replaced earlier concerns with *zhiliang* (which refers to a measured quality) and *wenhua shuiping* ('cultural level'), and resonated with time-honoured Confucian ideas of self-cultivation. It also echoed the Social-Darwinist concerns of nineteenth-century reformers like Liang Qichao, who believed that raising the moral quality of the Chinese people was essential to make the empire survive the threats of foreign imperialism. Sun Yatsen, the first Republican President, also shared this conviction (Sigley, 2009; Huang H., 2016), and Mao Zedong was a firm believer in raising the standards of the people, returning regularly to this concept in his writings (Murphy, 2004).

Despite the centrality of *suzhi* in the government's present narrative, it is extremely difficult to pinpoint the meaning of the term or what it entails in all its nuances, as it concerns cultural knowledge, ideology, politics, morality, and behaviour (Bakken, 1994). The task of raising the quality of the population of China individually and collectively is seen as central to the CCP's legitimacy in its attempts to produce a strong nation (Murphy, 2004; Kipnis, 2006). According to Kipnis, people use a perceived lack of quality to identify 'rural migrants, litterbugs, the short, the near-sighted and the poorly dressed' (2006: 296); in turn, Lin (2011) has established that people with low *suzhi* are brutal, impolite, retarded, self-abased, dirty, and dark. As a result of its multiple meanings, ascriptions, and usages, or its lack of semantic specificity in Chinese, the concept has been translated into 32 different terms in English (Kipnis, 2006; Sun, 2009). To put it shortly, *suzhi* has become a marker of social capital, a 'floating signifier', as Anagnost phrases it (2004: 197), with which one can define oneself or one's group in relation to others, to confer status and distinction (Sigley, 2009). *Suzhi* 'works ideologically as a regime of representation through which subjects recognize their positions within the larger social order and thereby sets up the conditions for socioeconomic striving' (Anagnost, 2004: 193). The term is 'highly mobile and seems capable of being deployed in almost any context in which comparisons between individuals, communities, and populations are being made regardless of gender, location, class, or ethnicity' (Sun, 2009: 618). According to Sun,

> urban understandings of low suzhi, of 'marginal places', of rurality and of the peasant and working classes, feed upon and reinforce one another. Consequently, all peasants are assumed to be lacking in suzhi, and anyone from 'the periphery' is viewed as peasant like (if not an actual peasant) and therefore lacking in suzhi (Sun, 2009: 629).

In having *suzhi*, one knows what to consume, thereby supporting the nation's economic growth and strengthening its global position, but one also cherishes and is aware of how to actively contribute to the maintenance of the social order that enables one's consuming behaviour; one is able to govern oneself in one's own community without the need of government intervention; and one constantly contrasts and compares one's social aspirations and behaviour against the models of individual self-improvement that are steadily presented in the media (Tomba, 2009: 592-593; Hu, Tu, and Wu, 2018). The present government continues to see the process of model emulation, the copying of the behaviour of others that was also embraced in Confucian, Republican, Maoist, and early Reform Era times, as the most effective mechanism for changing human behaviours (Munro, 1977; Blumenthal, 1979; Burch, 1979; Xing, 1984; Landsberger, 1993; Bakken, 1994; Lynteris, 2013). One can try and study, to learn or acquire *suzhi*. This is precisely what the many commercially available manuals and self-help books promise. The educational programmes that are offered to children and grown-ups alike, in both schools and the wider world, are another dimension of this self-improvement market. But essentially *suzhi* is a quality that one is born with, no matter how much effort one exerts to acquire it (Alpermann, 2013). There are many painful accounts of men and women from rural backgrounds doing their level best to 'blend in' in the new urban environments where they work, and yet failing miserably to overcome their identity as someone who does not belong (some are mentioned in Sun, 2009). The clothes they wear, the accents with which they speak Mandarin, their table manners and what they eat, their deportment – in the eyes of urbanites these are all dead giveaways of rural or non-urban backgrounds (Chew, 2003: 487).

Place-based stereotyping plays a major role in ascribing *suzhi*, both in the past and in the present, but such stereotypes are subject to change as a result of economic development and shifting power dynamics within the nation (Sun, 2009). Popular wisdom insists that people born in an urban environment have *suzhi*, meaning that those from the countryside do not. The *suzhi* of people born in tier-one cities like the metropolises of Beijing and Shanghai is of higher quality than the *suzhi* of the inhabitants of a tier-three or -four city, such as Lanzhou, the capital of Gansu Province. The urban district in which a person was born has a bearing on his or her perceived level of *suzhi*. Yet the perception of having or lacking *suzhi* also plays a role when evaluating the behaviour of people from the same social circle.

In the eyes of many, holding a socially acceptable and responsible attitude towards waste and garbage disposal has become closely tied to the concept of having human quality. According to this view, persons with high *suzhi* feel a responsibility to play an active role in maintaining the social order; and those should 'be blessed with high ethical and traditional values, be balanced in their relationship with neighbours, "speak truth and harmony", and have a strong sense of belonging' (Tomba, 2004: 602). They should not shrink back but instead speak out and show others the right way. Disposing correctly of waste and garbage, engaging in the separation of recyclables, and being seen doing so is part of performing one's *suzhi* – performing one's active role in maintaining the social order. Doing so gives one the authority to call out those who do not.

Knowledge and education – the central government

Judging by the number of policy decisions promulgated by the central government to curb the degradation of the environment and concerning the positive role of garbage separation and recycling by the population in this, the party-state is extremely concerned about these issues. Providing environmental education is therefore seen as a major responsibility for the government, in line with its long tradition of exposing the people to didactic stimuli. Earlier research has made clear that environmental education campaigns at the local level were initially lacking completely (Li S., 2002). This situation has changed over time. The 'Opinions of the Central Com-mittee of the Communist Party of China and the State Council on Further

Promoting the Development of Ecological Civilization' (Central Document Number 12), promulgated in 2015, for example, reaffirms the need for

> environmental education in schools, communities and government; giving 'full play to the role of news media'; promoting awareness about green policies and principles; fostering 'green lifestyles' among the public and officials through publicity campaigns, with an emphasis on 'thrifty, green and low-carbon' ways of living, and 'reporting advanced models' while 'exposing negative examples' (Geall, 2015c).

One of the problems the central authorities encounter is that their decisions and suggestions have to apply to the nation as a whole, forcing them to phrase them in the most broadly applicable terms. This, in turn, makes it difficult for the lower levels of government to see if, how, and where the central policies concern them, and how central policy directives can be turned into concrete rules and regulations.

This problem is evident in an analysis of the public service announcements related to garbage separation and recycling that have been produced over the years. First, such announcements are not broadcast very regularly, particularly when compared with the more politically oriented campaigns that are taking place at any given time. Even public service information related to climate change in general is broadcast more frequently than those regarding specific, concrete actions (Eberhardt, 2015). During the early stages of the campaign to mobilize the people for the Chinese Dream in 2012-2013, the environment in general and individual behaviour related to improving it played only a secondary role. Since President Xi Jinping launched the Chinese Dream programme after his election as General Secretary in November 2012 at the First Plenary Session of the 18th Chinese Communist Party (CCP) Congress, the ambition of 'realizing a prosperous and strong country, the rejuvenation of the nation and the well-being of the people' has figured prominently in government communications (Wang, 2013; Bislev, 2015; Callahan, 2015). Over time, more concrete definitions of the Chinese Dream have emerged that make these goals clearer. The Dream consists of four parts: a Strong China (economically, politically, diplomatically, scientifically, and militarily); a Civilized China (equity and fairness, rich culture, high morals); a Harmonious China (amity among social classes); and a Beautiful China (healthy environment, low pollution) (Kuhn, 2013; Landsberger, 2014, 2018). Framed in terms of raising (human) quality, state educators and public intellectuals have made the education of the people concerning the Dream's morals and values one of their major responsibilities. The educational materials that have been produced in the process and the frequency with which the goals are addressed makes it clear that the ranking of these four parts also indicates their relative importance (Fieldnotes 2015, 2017). Hence, a healthy environment and low pollution rank last.

The materials the government has designed to educate the people in the Chinese Dream do not provide any concrete learning tools and goals when it comes to the environment (China Civilization Office, 2013; Fieldnotes 2015). They suggest a desired behaviour, rather than show it. There are posters that call for economizing on the use of water and fossil energy; that call on people to use less disposable food packaging; to order food with moderation so as to cut back on food waste; to take bicycles instead of driving cars (this predated the fashionable use of shared bicycles that peaked in 2017, as seen in Chapter 2); they call for a reduction in the use of paper, etc. Some of these propaganda and educational materials, ranging from posters that are put up in the streets to desktop images that can be used on computers, are original and stunning and have been produced with high production values. But they lack concrete suggestions or behavioural models that can be put to immediate use. Most of the Chinese Dream propaganda that was produced after 2013, including public service announcements made for broadcast on television, focused on the natural beauty of the nation, not on ways to preserve it. Only a very few dealt explicitly with things that concerned ordinary people. This included one message that was produced by the Wuxi, Jiangsu Province, Propaganda Bureau in 2014. The short video clip explicitly linked the garbage separation behaviour of a young couple with the wife's desire to leave a better world for their future children. To be more specific, the wife dumps the separated garbage into the designated bins and lectures her husband while doing so ('Waste sorting', 2014). This reflects the gendered practice of waste disposal discussed in Chapter 3.

A departure from the general rather vague educational clips was the release of an online video animation entitled 'Xi Jinping cares about these six things' in 2017, shortly before the convening of the 19th CCP Party Congress (Beijing Municipal Jingmeng Culture Communication Company, 2017). Accompanied by an upbeat rap tune, it addresses six issues that are important in people's daily lives, and states that Xi Jinping also cares about these very much. These issues are air pollution (in particular PM2.5 pollution[16]), garbage recycling, bio-waste pollution, food safety, care for the elderly, and

16   PM2.5 refers to atmospheric particulate matter (PM) that have a diameter of less than 2.5 micrometers.

housing prices. For each of these problems, the video suggests a quick and seemingly easy solution. The catchiness of the tune, the rap lyrics, and the way in which the video is produced all suggest that millennials, or young(er) people in general, are its target audience. The younger generations, as is widely acknowledged, need to be told how to behave. But the main rationale behind such messages is to show the people that the government shares their concerns and is developing policies to abate the problems. In other words, the government is taking action; the people do not have to worry.

Knowledge and education – the municipal government

When it comes to defining what responsible behaviour with regards to waste separation and recycling is, in Beijing almost everyone has his or her own opinion and ideas. This is caused by the fact that different regimes of authority make use of different standards; the central municipal government, for example, leaves the district governments with considerable leeway in how to define such behaviour, and in turn the district governments leave a lot of the responsibility for educational work to the residential community committees. It is quite likely that this situation exists in the rest of urban China as well, but that may not necessarily be the case. One can assume – and this view was corroborated in a number of talks with municipal representatives – that smaller cities are better equipped to provide their inhabitants with relevant information, since the lines of communication are shorter and the provision of information can take place in more personalized settings. Moreover, within these communities, peer pressure is stronger and more able to change behaviours.

Ms. Zhang, a staff member of the Solid Waste Management Office of the Beijing Municipal Appearance Management Committee under the Beijing Municipal Commission of City Management, who is responsible for educational projects and materials, indicated that the activities they organize often take place in isolation: they do not take place on a city-wide scale, or as part of a national drive; they are one-off events and not parts of recurring and broader campaigns; and there is not much room for evaluating the efforts, exchanging experiences, or drawing lessons from them. In my talk with Ms. Zhang, she stated that the issue of garbage was of paramount importance; municipal education and propaganda projects should take the task of spreading knowledge of the basic rules of garbage and garbage disposal as a priority, and citizens should aim for the reduction, reclamation, and disposal of their harmless waste. These objectives formed part of the broader municipal social civilization project and reflected the concerns expressed at the central levels of government (Heberer and Senz, 2011). To achieve the aims of these educational projects, it was important to make the citizens understand the meaning and use of garbage classification and reclamation. Such basic knowledge should help to make them understand that everybody could and should play a role in improving the environment and reducing consumption. In the end, it is most important to let everyone realize that they should change their habits and behaviours.

The Solid Waste Management Office has organized several projects and events in the recent past. Since 2010, every Thursday had been designated 'Garbage Reduction Day', while the last Saturday of every month is marked as 'Renewable Resources Day'. Although other sources have suggested that the introduction of such days was an eminent opportunity to organize awareness-raising activities and provide knowledge (e.g., interview with Hong Chao, 2017), I was unable to find out which special activities took place on these designated days, where they took place, and who participated in them. These days could have been examples of symbolic efforts. For a number of years the Office also has organized 'Garbage Culture One-day Tours' on every Thursday, targeted at residents, primary- and secondary-school pupils, and university students. On such tours, the participants visit garbage incinerator plants such as the Gao'antun incinerator facility introduced and discussed in Chapter 7; parks that have been built on converted landfills of construction rubble and other waste; and enterprises reusing plastic bottles, like the Incom Company introduced in Chapter 2. The more visibility and publicity that these events are able to generate, the easier it is for the Solid Waste Management Office to demonstrate that its work is in line with the policy directives coming from higher administrative levels (Heberer and Senz, 2011).

Green Frog

Ms. Zhang proudly referred to her personal involvement in the writing and editing of several of the Office's publications. One in particular, *The Story of Garbage* (垃圾的故事), was recommended nationally as one of the 'Top 100 Good Books for Young People' in 2015 by the State Administration of Press, Publication, Radio, Film and Television, the state agency that oversees the media. Moreover, the China Environmental Science Association awarded the book, which is targeted at an 11-14-year-old readership, with the Environmental Creativity Prize in 2016 (Office of the Capital Spiritual Civilization

Construction Committee, Beijing Municipal Appearance Management Committee, 2010b). Such accolades are important not only for the Solid Waste Management Office but also for the individuals concerned, because they have a positive effect on work evaluations (Heberer and Senz, 2011). Other educational activities in which the office was involved included the production of educational videos and cartoons, posters, and smartphone applications to facilitate garbage classification. The videos and cartoons are broadcast on television and on the small TV screens in subway cars and other forms of public transport (Eberhardt, 2015). Moreover, the office organized and hosted several well-attended public lectures over the years.

A closer look at some of the materials published by the Solid Waste Management Office shows that they favour the use of cartoon images to spread garbage-related information (Office of the Capital Spiritual Civilization Construction Committee, Beijing Municipal Appearance Management Committee Office, 2010a; 2010b; 2014). These cartoon figures are already known to the public, such as the Green Frog that features in *Green Frog in Action* (Office of the Capital Spiritual Civilization Construction Committee, Beijing Municipal Appearance Management Committee, 2010a). This particular frog has appeared in many public service messages on television and in the subway systems; its recurring use makes easy recognition and internalization of the messages possible. In the *Household Waste Classification Guidebook for Citizens of the Capital,* a publication aimed at a more mature audience, the Green Frog has been replaced by a cartoon waste bin (Office of the Capital Spiritual Civilization Construction Committee, Beijing Municipal Appearance Management Committee, 2014). The *Guidebook* presents the most relevant and essential information through illustrations of recognizable family groupings and situations, with parents, grandparents, and single children. The children are the ones towards whom the message is directed; the adults usually perform the behaviour that needs to be corrected and who need the guiding hands of the children. The adults can be divided in two groups. The ones displaying the behaviour that needs to be changed are invariably the parents; the elderly, i.e., the grandparents do not need any instructions, as it is clear that they know what to do. The focus on children and young adults closely follows the broader philosophy that guides these propaganda efforts; according to Ms. Zhang: 'We witness the impressive influence of "small hands holding big hands". One child can influence three generations and the whole family. If we have confidence in the next generation and the future, our dream of environmental friendliness will come true and society will achieve a resources-saving behaviour and environmental friendliness' (Interview with Zhang, 2017; Beijing Municipal

Urban Management Committee and Beijing Municipal Urban Management Committee Information Center, 2016:4). Using literature and art to educate children and young people is not a new approach; they have been used as ideological tools by all Chinese political regimes that have been in power since the beginning of the twentieth century (Blumenthal, 1979; Farquhar, 1999; Donald, 1999).

Moreover, the Solid Waste Management Office actively uses social media such as Weibo and WeChat to spread information and propaganda and to collect ideas about how to further spread knowledge about recycling. In cooperation with the Beijing TV (BTV) Station Life Channel,[17] the Office has also produced special TV programmes, created news items, and organized various online competitions for viewers, all with an educational objective in mind. The Office often teams up with other administrative bodies for these media productions (e.g., Beijing Municipal Statistics Bureau, 2017). The Office furthermore is responsible for publishing relevant print publications. One of these is the bimonthly magazine *Urban Management and Science and Technology* (城市管理与科技),[18] which, amongst other things, regularly reports on environmental and garbage-related educational activities in residential communities in its section 'Communication Platform'. Based on the content listing on its website, however, the journal seems to have ceased publication after Issue Five of 2017.

Knowledge and education – urban drives

The reports in *Urban Management and Science and Technology* give an indication of the formats that such information-provision events or drives generally adopt. They follow a script that is not restricted to environmental topics but that plays out regularly on the streets of urban China (Boland and Zhu, 2012). Drives are the prime opportunity for health institutes to organize blood donation drives; for legal departments to educate the public about changing rules and legislation; for government offices to publicize new policies, and so on (Fieldnotes, 2015, 2017). Such a script unfolds like this: on a Saturday (sometimes Sunday) morning or afternoon, a stage is erected in a community (or in a school yard, on the sidewalk of a crowded street, in a shopping mall, or in a square) and a display screen is put up with information about the activity that will take place. A sound system is put in

17   https://www.youtube.com/channel/UCaMJ1qzUDgev9N1ye_uKJoA
18   Issues are partially available at http://www.bjmac.gov.cn/sy/syztzl/M/

place, and tested; popular (hit) music is played at a high volume to attract the attention of passers-by. A number of representatives from the municipality and the community (or the school, the department, the hospital) mount the stage to give speeches in which they express their own or their organization's commitment to whatever needs to be supported. Volunteers wearing special tee-shirts and ball caps mingle with the bystanders or sit behind trestle tables; they hand out information brochures, leaflets, or trinkets, or engage in face-to-face education. Bystanders and passers-by throng around these activities out of curiosity; there is a lot of camaraderie and fun; there may be games and other special child-related activities planned – in short, it is a special day for all involved (Fieldnotes 2015, 2017). Although many people participate and thus receive the message of the day, these are all one-off events. They generate a lot of (media) attention, lots of photographs are taken that can later be used in publicity materials to illustrate how committed the organizers are. But usually these single events are not followed up, and the information that has been provided is not hammered home on consecutive occasions that can build upon the initial effort and nurture a lasting commitment (Simões, 2016).

The planning and preparation behind such drives is clearly visible in a plan that was drawn up jointly by the Office of the Capital Spiritual Civilization Construction Committee and the Beijing Municipal Appearance Management Committee in April 2010. Entitled the '"Be a Polite Beijinger, Reduce the Amount of Garbage and Start Garbage Classification by Yourself" Plan', it set out a number of activities that the two administrative bodies were going to organize over a period of eight months (May-December 2010) (Office of the Capital Spiritual Civilization Construction Committee, Beijing Municipal Appearance Management Committee, 2010c; Johnson, 2013a). This campaign coincided with the designation of 600 pilot communities to start garbage separation; 30 percent of the government offices and institutions as well as schools also received the order to participate in the drive. It remains unclear whether these offices and schools were located within the pilot communities, or were included as complementary pilots. Aside from encouraging garbage separation, the aims of the campaign were manifold: to start a 'zero-waste' trial period in 100 institutions (not further specified); to reduce the amount of waste produced; to turn the waste into reusable resources as much as possible or dispose of it as harmless waste; and to control the rate of the increase of garbage in the city to about 5 percent. Various sectors of Beijing urban society were targeted during the eight months of the campaign.

May 2010 was named Green Office Month, which mainly focused on Party and government offices, institutions, and companies. These were urged to cut their number of meetings and the documents used, and to promote the use of electronic documents and services/devices. Participating offices were ordered to save water, electricity, and energy, and encouraged to use double-printed paper, recyclable paper, or less paper overall. June 2010 was made into the Green Food and Beverage Month, which focused on the residents. They were told to only order the amount of food that they could actually consume, and to take the leftovers home. Restaurants were instructed to use fewer disposable eating utensils and to display 'green food and beverage' advertisements. The specific goal of this month was to reduce garbage at the front-end. The tourist industry was the target in Green Tourism Month, held in July 2010. Organizations and units active in the tourist trade had to design slogans and advertisements that promoted garbage reduction, garbage classification, and zero waste. They were also instructed to encourage tourists to bring their own food, snacks, and drinks, and to use fewer disposable products. The tourist companies were told to collect all of the green waste and turn it into resources on the spot as much as possible. In August 2010, attention shifted to creating Green Living District awareness among the residents. For this, lectures about appropriate topics and tours to relevant destinations needed to be organized. Residents had to be made aware through the screening of relevant movies, public lectures, and posters. Students living in the participating districts or communities also had to be mobilized and motivated to promote the concepts of the campaign during their vacation. September 2010 was Green Business Month. Shops and businesses had to be mobilized to use less packaging (materials), to offer clean products, and to promote the circular economy as much as possible. Vegetable markets were specific points of attention; vegetable and fruit sellers were told to only sell produce that had been cleaned, i.e., with the stalks and leaves taken off as much as possible. The issue of clean products was revisited in October 2010, during Green Market Month. The peasants and business owners who sold goods in rural markets had to be encouraged to use fewer plastic bags and sell clean vegetables. Fruit and vegetable waste had to be disposed of on the spot or in the nearest garbage dumping facility. Attention turned to universities and other educational institutions in November, for the Green Campus Month. This meant that activities were organized there, consisting of providing information on blackboards, the publication and distribution of relevant magazines, and the tapping of school networks. Ideally, this month would offer the opportunity to build volunteer teams that could be mobilized for future events. Finally, in December 2010, the experiences of all the involved units needed to be shared and summarized.

This plan only set the themes and broadly suggested the types of activities that needed to be organized during each of these time periods. It left the actual implementation to the lower levels that would have to do the actual work. At the time the plan was adopted, no clues were given about how these activities might be realized, no suggestions were offered for follow-up activities, and no preliminary plans were made for the following year(s). Despite the provision that experiences would be summarized in December 2010, no information has been made available about the end results of this plan.

Knowledge and education – online resources

Information regarding garbage, garbage disposal, and waste separation is obviously available to residents in large quantities and a wide variety. The municipal government and district governments in Beijing all have websites that provide detailed information on these and related topics for whoever is interested.[19] The Leading Group for the Implementation of the Propaganda Movement 'Be a civilized and polite Beijinger – waste reduction and garbage classification start with me' under the Office for the Creation of the Capital's Spiritual Civilization Committee and the Beijing Municipal Appearance Management Committee provides relevant and regularly updated information.[20] Many branches of the solid waste management industry also run expansive websites providing detailed information.[21] It is often a matter of knowing where to look to find the sites and the information one needs. This is time-consuming and thus not efficient. Many if not most of the websites that have been put up by such government offices and institutions have not been updated for a long time (sometimes the information provided is a few years old), contain information that is no longer relevant, present the user with a 404-error message, or have gone offline completely. One almost gets the impression that these departments have only gone through the motions of complying and engaging in a desired behaviour (i.e., creating an informational website) to satisfy higher levels of the government, as some of my sources have suggested. Setting up a topical website serves as an illustration of departmental compliance. It certainly is

19 Extended information about garbage disposal, garbage separation, and recycling as provided by the Beijing Municipal Commission of City Management can be found here: http://www.bjmac.gov.cn/hjwsbz/.
20 http://zt.bjwmb.gov.cn/ljjlfl/
21 For example, http://www.solidwaste.com.cn/.

not a lack of information that is keeping ordinary residents from engaging in behaviours that are beneficial to the environment, as my resident sources claimed. Instead, not engaging in the desired behaviour desired seems to be a personal choice. Moreover, it seems that the information residents really want in order to make informed decisions is not what is available to them.

Aside from these official information sources provided by governments at various levels, most of the (semi-) private companies that are involved in garbage disposal, retrieval, and recycling in one way or another run websites containing information that they think is relevant and beneficial for citizens (as well as for their own operations). O2O companies like Incom, Taoqibao, and Zaishenghuo regularly publish new information on their websites, and use their Weibo or WeChat accounts to mobilize their subscribers or clients to persevere in environmentally friendly behaviours. Many of these messages use national and/or traditional holidays and events as markers for their information, as these moments generally involve various forms and types of gift-giving or call for the large-scale consumption of traditional foodstuffs – all of which results in an increase in garbage. The Qingming Festival or Tomb Sweeping Day, for example, takes place every year on 5 April. On this day, families visit the graves of deceased family members and relatives to 'sweep' them and leave funerary gifts behind to be used in the afterlife (Illustration 5.1). A visit to the ancestral grave is not complete without setting off fireworks (prohibited in most urban areas these days), burning incense, and leaving behind (paper) flowers, papercuts, and garlands, as well as edible and ceremonial foodstuffs. These and other activities lead to an increase in waste on top of what is already discarded on a regular basis.

The yearly Singles' Day extravaganza on 11 November is another example of almost orgiastic urban consumption. Singles' Day is just one of many other comparable retailer-driven occasions that take place every year in China. The number 11, the two digits that mark both the day and the month, resemble bare sticks or branches, which is another way of identifying single people. The Internet retail giant Alibaba started to promote this event in 2008 as a Chinese alternative to Valentine's Day, urging people to think of poor singles and make their loneliness more bearable by buying them a gift. Over the years, it has grown in popularity among consumers, as well as in importance for manufacturers and retailers. In the weeks leading up to 11 November, potential consumers are bombarded with advertising stimuli on television, their smartphones, and other media. During the one-day event itself, people buy goods using their smartphones (and credit cards) almost compulsively, not because they might need these goods or have a use for them but because they have been mesmerized by the discounts they

Illustration 5.1   Overflowing waste bins at a Beijing cemetery during the 2017 Qingming Festival, or Tomb sweeping day

Author's photograph, 4 April 2017

have been promised. If the promises of discounts do not succeed in hauling in buyers, the free gifts they are offered in the course of their extended online shopping behaviour most certainly will. On 11 November 2017, the number of items ordered was estimated at one billion pieces (Walsh, 2017). The environmental impact of Singles' Day is not only that people buy many unnecessary items, which all have to be delivered by courier services. Upon opening the packages they have bought for themselves or others, people are also faced with mountains of paper, plastic, styrofoam, and sticky tape, as their deliveries are packaged in ludicrous heaps of packaging materials. And these packaging materials need to be disposed of, despite reassuring noises from the retailers that they are working on the introduction of recyclable, multi-use forms of packaging materials, or reductions in the size of the packaging or packing tape (Gupta, 2017). As past experience has proven, most of the packaging material ends up in the ordinary waste stream, rather than in the waste bins intended for separation.

Apparently part of a concerted customer relations management strategy, the O2O companies update their subscribers/clients at least once a week about company news and activities, new services, semi-personalized offers, etc. The underlying philosophy of these communications is to incentivize garbage disposal and recycling, and to show that it is worth people's time to participate in the companies' programmes. However, in the spring of 2018, only the Huishouge Company was still active in providing this type of information; the competition seems to have fallen silent. The users of the apps rarely look at the information the companies provide, and new users do not go to the trouble of scrolling back to earlier messages (Interview with Friends of Nature, 2017).

In the community

Even without searching for specific information about garbage separation and recycling on government websites or the sites run by O2O recycling companies, it is almost impossible not to be informed of the basics of these practices. Almost any regular newspaper pays attention to these topics as a matter of course, such as, for example, the very informative 'How should we recycle' (垃圾分类，我们该怎么做) article that appeared in the *Beijing Evening News* (2017). This article gives an overview of Chinese practices that are presently embraced and devotes attention to experiences from abroad that can be emulated, particularly those from Japan.

Many of the residents I spoke with, especially the younger people who work regular jobs (i.e., those in the age group of 25-40 years), indicated that, as far as they could remember, they had never received any kind of education or information about garbage classification and separation while in school. For a long time, environmental concerns have taken a back seat in educational practices. In the Maoist era, nature had to be harnessed and exploited by humanity (Shapiro, 2001); in the early Reform Era, post 1978, economic development received most attention. In 1973, the first steps were taken to develop an early form of environmental education, although it mainly took place at university levels. It was only in 1992 that pilot programmes were established in kindergartens and at the primary and secondary school levels (Wu, 2002; Tian and Wang, 2016). This date explains why the people I talked to claimed they had not received education on this topic in school. Some of the residents maintained that they had not encountered or participated in any of the special environmental or garbage-related events that took place in their communities, nor had they seen any campaign materials published during such environmental drives. Others certainly were aware that environmental drives were taking place in their communities, with posters and slogans in the yards put up by their communities' management committees – or at least they said they were. Yet none of these promptings had been able to increase their knowledge, change their behaviours, or improve their willingness to act. On a side note, I should state that almost all of the people I spoke with saw environmental awareness and garbage recycling as one and the same thing: recycling one's garbage meant that one was environmentally aware, and vice versa.

Of course, many respondents claimed that they separated their garbage into the three (sometimes four) different waste containers that were generally provided in their communities, although this self-reported behaviour could not be tested (Tang, Chen, and Luo, 2011). But at the same time, many others professed their utter ignorance about separation and claimed they did not recognize the meaning of the identification symbols stuck to the garbage containers in front of their building or in the yard. Likewise, when it came to identifying which piece of garbage fell into exactly which category and belonged in which container, even the people who were proud of their separating and disposing behaviour admitted that they often simply did not know. The classification of food containers is as an example of their uncertainty: most people consider them to be food waste and discard them as such, whereas the containers should actually be disposed of in the bins for recyclable plastics. Many community residents confessed that they often chose the line of least resistance by putting their garbage in the container with the open lid, no matter whether the garbage matched the classification on the bin. A similar tendency was also mentioned in the extensive write-up of the garbage pilot project that the ENGO Friends of Nature (FoN) produced in 2013. As a last resort, there is always the 'guy downstairs' to help people sort out their garbage.

It should be obvious at this point that it is not a lack of information, or of the comprehensiveness of the information, that keeps people from engaging in garbage separation. The fact that the information provided does not make people commit to changing their behaviour rules out the possibility that they will do so. This raises questions about the effectiveness of the education-by-role models that the party-state continues to see as the most successful method of instruction. This belief assumes that continuous and endless confrontation with a desired behaviour will almost automatically lead to behavioural change. But how much continuous and endless education is needed to bring about this change? As Goldsmith and Goldsmith (2011: 120) have already noted, merely using pamphlets or lectures does not lead to more environmentally friendly behaviour. Additionally, it has become clear that most of the environmental information that is supplied is couched in the most general terms possible and does not necessarily answer the concrete questions that continue to bedevil people. Simões (2016) argues that it is necessary to resort to other types of mechanisms to beyond the well-tested public information campaigns; Tang, Chen, and Luo (2011: 856) suggest paying more attention to the effects of social pressure and of instilling an awareness of the moral correctness of a specific behaviour.

It is clear from the various policies adopted over time by the various levels of national and municipal government that a uniform classification scheme of waste has still not been formulated in China, let alone been published or enforced. As before, recently adopted legislation explicitly leaves the responsibility for defining these categories of waste to the local governments (State Council General Office, 2017). While allowing local policies to be formulated on the basis of local customs or locally available disposal or recycling opportunities makes some sense, at the same time it contributes to the confusion. What is seen as a recyclable good in one locality is incinerated in another.

Turning information into concrete behaviour

Few of the promptings to dispose of garbage or recyclable waste are taken to heart by urbanites. The residents listen to the exhortations, receive the leaflets that are handed out, and maybe study them thoroughly, maybe even pledge to participate. But in the end, as various pilot projects (Friends of Nature, 2013; Yuan and Yabe, 2014; Dai et al., 2015) and my own interviews have shown, this newly adopted behaviour is not completely internalized, and slippage occurs over time. The administrative bodies that put out these promptings are still convinced that merely deciding on a course of action is enough to generate change. Once they have done so, the people will simply participate (Teets, 2013: 11). What is obvious from these pilot projects and interviews is that a comprehensive change of behaviour can only be accomplished when residents are constantly corrected, almost taken by the hand by others who will evaluate their behaviours and try to improve them by providing concrete examples. Many urbanites continue to hold on to the idea that whatever they do, it does not make any difference; that their individual acts are unable to change anything. Some city dwellers indicate that it is the government's responsibility to deal with the garbage problem. They are the ones who pay fees to make it go away, after all, so now it is up to the authorities to act. And many maintain that if others do not change their behaviour and the government does not enforce it, there is no point spending any time or effort on it themselves.

Yuan and Yabe (2014) reported that, by the end of 2012, 2412 residential communities in the Haidian and Dongcheng Districts had become involved in the pilot projects devoted to garbage separation that had started in 2010. In a number of cases, these projects included the separation of kitchen waste as well. To make these experiments work, the local governments had to hire 'garbage separation instructors' (each responsible for 80-120 households) to provide information and guidance. The ENGO Friends of Nature analysed the effects of these pilot projects in 60 of the communities that were involved. They concluded that only a few results had been accomplished. While the separation rate had improved from 5 percent before the pilot started to 15-20 percent, Friends of Nature had expected that the hands-on management of the separation campaign would have yielded an improvement of 50-60 percent (Friends of Nature, 2013). To their great consternation, moreover, when they later returned to the pilot communities for a follow-up analysis, the Friends of Nature volunteers discovered that the separation percentage had again dropped, almost to the levels that existed before the trial started (Interview Friends of Nature, 2017).

In fact, the garbage separation instructors that were hired by the sanitation departments or the management companies to keep an eye on the work and at the same time provide continuous education among the residents, were forced to act as actual garbage separators themselves in the end. Many residents were convinced that the instructors, some sporting green badges, others dressed in green work clothes or wearing green sleeve protectors (earning them the nickname 'greensleeves'), were there to do the actual separation work for them. In a number of communities, the local governments were forced to restructure the contracts with these 'greensleeves' and employ them as secondary sorters to resort the garbage. Some of these teams of 'greensleeves' decided to sell the large quantities of recyclables they had separated on their own and pocket the money (PKU MBA Deep Dive, 2015). The laudatory articles that were published by the municipality to pay tribute to the contributions of these workers (see Cheng, 2012; Zhang, 2017) do not report on these events. Instead, these write-ups rather uniformly describe the honour the instructors felt to be hired for this task and the harmonious relations that emerged between them and the regular residents.

As Goldsmith and Goldsmith (2011) made clear, a more personalized, hands-on effort to persuade people to change showed a positive effect in other experiments as well. Dai et al. (2015) undertook a doorstepping project in residential communities in Shanghai to monitor food waste separation. Friends of Nature has organized a similar activity in Beijing (Interview Friends of Nature, 2017). Dai et al. discovered that engaging the targets in one-on-one communication with volunteers who interacted with the residents by visiting them, making clear that each individual's contributions did matter, had a significantly positive effect on capture rates (an increase from 45.2 percent to 57.7 percent). Aside from knocking on residents' doors, the volunteers also provided hands-on sorting assistance at the garbage bins in the yards. These volunteers were young people, students from nearby universities, who displayed a remarkable enthusiasm in their interactions with the residents. This youthful commitment by budding intellectuals had a significant influence on the people on whose doors they knocked. The residents were impressed by the trouble these volunteers went through, admired their stamina, and complimented them on their work, even though they were not part of the community. The residents felt that there was increased social pressure to conform; at the same time, the volunteers were seen as having a high level of *suzhi*. By following the prompts of the volunteers, some of their perceived higher quality would be transferred to the residents. Yet the general feeling among the inhabitants remained that the effects of this doorstepping would not last. They were convinced that they needed to be reminded every few months of their duty to separate their garbage, either by student volunteers or representatives of the community management committee or the real estate agency (Dai et al., 2015: 9, 11, 15-16; Interview Friends of Nature, 2017).

Simões (2016) suggests that in the end, market-based instruments will yield the best results by appealing to people's self-interest. These should be based on a set of well-crafted incentives and disincentives, ranging from subsidies to fines. Other research supports this (Tang, Chen, and Luo, 2011: 869).

What do the people say for themselves?

The Friends of Nature report (2013) provides a treasure trove of information and insights about residents' attitudes. The suggestions of the residents themselves about how to improve their awareness of the garbage problem are of particular interest. Around one half of the residents who were questioned suggested that consistent propaganda and education drives and efforts by the authorities (residential community committees, community management companies and real estate agencies, district and municipal governments), especially aimed towards children and the young, would yield favourable results. The suggestion that these age cohorts needed special attention indicates that they themselves already knew everything that was needed. 30 percent of those polled were convinced that handing out awards and certificates to those who displayed desired behaviour would motivate other residents in a positive way. They were also convinced that more lectures would yield an improvement in behaviour, especially when they were organized in conjunction with poster campaigns (see also Teets, 2013). Some residents also proposed that more strict measures against violators would improve the situation and believed that special propaganda efforts directed at newcomers (i.e., leaseholders) would help them understand what was expected of them (Friends of Nature, 2013). This distinction between original residents and newcomers involved assumptions about quality, which the latter were accused of lacking. As these newcomers had not been part of the original cohort of residents, they were seen as different, as the 'other', and therefore were singled out by the original residents for special attention.

Turning behaviour into positive credits

Since 2014, the government has been working on the introduction of a Social Credit System that will make it possible to rate the individual behaviour of members of the Chinese public (State Council, 2014). The details of how this System will work and how credits will be awarded or deducted are still shrouded in secrecy, but in general the system is presented as a tool to help improve the quality of the population. In the words of Rogier Creemers,

> it is intended that social credit information will be connected with individuals' identity card numbers, creating unique and traceable files that can be used to facilitate citizens' access to financial and government services. At the same time, the plan called for the introduction of real-name identity-based appraisal and scoring of individual online comportment, as well as of blacklists for those perpetrating various kinds of fraud, deception and 'harm to others' lawful rights and interests'. (2017: 97).

According to Shi-Kupfer et al., the System will collect

> online data from individuals and companies to track and influence not only economic honesty and credibility, but also social and potentially political behaviour. The Social Credit System includes incentives for behaviour defined by the CCP as 'good' (rule-conforming) and disadvantages or outright sanctions for 'bad' (rule-breaking) behaviour (2017: 18).

In its present form, the System has an economic component, which will serve as an effective social and economic management mechanism to establish,

among other things, a person's creditworthiness, thereby lowering operational costs for companies and improving the commercial environment. The
second component deals with the need to increase transparency and create
broader channels for public participation in government decision-making
(State Council, 2014), but it is unclear how the Social Credit System will
be able to contribute to this on the level of the individual. The third part
is the establishment of judicial credibility, defined as the 'credibility of
the courts' and the extent to which judicial processes are considered to
be in accordance with the law (State Council, 2014). As with the previous
component, the potential effects of the System in this field remain unclear.
Most important for this discussion, however, is the fourth component, which
will rate a person's social creditworthiness. Under this rubric, a person's
ratings are calculated by rewarding trustworthiness and punishing the
reverse, raising the trust level of the entire society in the process. This
element of social credit in particular has dominated Western reporting
on the System, largely focusing on privacy concerns and assumptions that
its introduction will turn China into a state reminiscent of that in George
Orwell's dystopian *1984* (Greenfield, 2018).

To prepare for a smooth, nation-wide introduction of the system, a number
of pilot projects have been set up in various cities for testing. The trial that
took place in Hangzhou, Zhejiang Province, was awarded the first national
award for the development of the Social Credit System for its accomplish-
ments. The Hangzhou credit score is based on an elaborate report that
comprises basic information (information linked to the formal identity
and identity card of a person), positive information (related to social ethics,
professional ethics, family virtues, and personal qualities), and negative
information (ranging from serious crimes to petty offenses, such as parking
fees and jaywalking), and other information. Factors that can have a positive
effect on one's score include doing volunteer work, participating in blood
donation drives, waste sorting, and 'other social services' (Zhuang, Wu, and
Jia, 2018). If the Social Credit System, once it is rolled out, will indeed add
garbage classification, separation, and related activities to the rubrics that
have a positive influence on one's credit score, this will fit with the desire
felt by many to reward environmentally conscious behaviour. Linking it
to potential rewards under the System may sway more people towards
cooperation. Technologically speaking, it will not be difficult to do so. The applications that have been developed by Incom (Bangdaojia) and ZaiShenghuo
already collect and add all information about what an individual offers for
recycling, keeping a running total. These data can presumably be integrated
into the broader framework of the Social Credit System.

6   NGOs and other voluntary environmental groups

Voluntary environmental organizations are very active when it comes to developing activities aimed at raising the consciousness of the population concerning garbage disposal, garbage classification and separation, and the benefits these activities have for improving the living environment. International organizations like Greenpeace International and domestic ones such as Friends of Nature, Green Beagle, and (many) others are taking the lead in these fields in urban China. All non-governmental organizations (NGOs) have to fight a battle on two fronts, particularly when they concern themselves with causes that can be construed as having a political impact. As environment-related causes tend to be intimately linked with political concerns, this also applies to those NGOs that are working in the field of the environment (ENGOs). Aside from encountering interference while engaging in their core business of educational and activist work, they also face continuous and stiff obstruction from official quarters to be active at all. The Chinese party-state considers organizations like NGOs as a threat to its existence and monopoly on power. This seems to make the work of these organizations extremely difficult, but practice has shown that many of them have still found ways to engage in relevant action. NGOs generally need to work in the background as much as they can; they must try and avoid too much publicity for their actions; they must ensure that their activities carry no political implications; they have to arrange for financial resources while avoiding funding from abroad, etc. At the same time and despite such constraints, some NGOs are able to successfully negotiate the divide between the people and the political system. Some are even appreciated by that system, albeit incidentally and grudgingly, for their contributions (Lu, 2007; Salmenkari, 2008; Wu and Chan, 2012; Johnson, 2013a; Kostka, 2014; Teets, 2014). As Dai and Spires conclude, over time some ENGOs have been able to carve out a position 'as watchdogs to government policies, calling for implementation of existing regulations, critiquing and campaigning against undesirable policies, and exerting pressure on government to solve environmental problems' (2017: 63). Others have provided invaluable assistance in the implementation of government initiatives by going into communities, linking up with the residents, and mobilizing support – activities that the government bodies are not able or willing to do.

Registration

The phenomenon of the environmental NGO made its appearance in China fairly recently. The first independent environmental NGO, now known as Friends of Nature (FoN), was officially registered in 1994, following the unsuccessful Chinese bid to host the 2000 Olympic Games. The fact that China did not have NGOs at the time of the bidding process is said to have contributed to its loss. FoN was set up by the late Liang Congjie, the grandson of Liang Qichao, a reformer during the waning days of the Empire. He was also the son of Liang Sicheng, an architect who tried to preserve the historic city walls of Beijing as much as possible when the new government wanted to tear them down and turn the city into a showcase for socialism after 1949 (Larson, 2010; Tsang and Lee, 2013). Moreover, Liang Congjie served as a member of the Chinese People's Political Consultative Conference, the highest advisory body for the government (Schwartz, 2004). Liang's commitment to environmental advocacy resulted from his awareness that it was one of the causes around which it was safe to organize. An effort to protect the Tibetan antelope from poachers organized by FoN in the 1990s proved successful; campaigns to stop dam construction in the Nujiang River (Yunnan Province) and to preserve a scenic section of the Yangtze River called the Tiger Leaping Gorge, organized in the 2000s, also ended in victory. In 2004, former Premier Wen Jiabao personally stepped in to put the Nujiang dam plans on hold (Larson, 2010).

Despite Liang Congjie's stellar pedigree and political connections, the registration process of his organization was not without problems, as Tony Saich recounts: after waiting for ten months upon handing in their request for sponsorship, FoN received a reply from the National Environment Protection Agency, the administrative level that was to serve as their sponsor. Liang learned that his organization could only be registered if it would take on the responsibility of representing the interests of all Chinese who shared its environmental concerns. Liang declined these demands and instead registered in 1994 as a secondary organization with the Academy of Chinese Culture, the institution where he served as a professor and vice-president (Saich, 2000: 131, 138). In the wake of FoN's recognition, many other environmental NGOs became active, including Global Village of Beijing. The Bureau for the Administration of Non-Governmental Organizations under the Ministry of Civil Affairs, is the organization responsible for formal registration. Being formally registered is extremely important for an NGO to be able to function properly. The most preferred registration status for NGOs is that of 'social organizations with tax-exemption status', but in every local jurisdiction only one social organization is allowed to obtain such status for the theme in which it is active. At the

Beijing municipal level, for example, this means that only one environmental protection association in the entire city can be granted approval (Wu and Chan, 2012: 10). To register for that status, an NGO must find a sponsor organization that is either a governmental agency or a government-affiliated organization. It is not allowed to set up branch offices and has to submit an annual financial report (Schwartz, 2004; Hildebrandt, 2011; Wu and Chan, 2012; Zhan and Tang, 2013: 385). Non-registration means that the NGO does not occupy a niche within the existing administrative framework. In the eyes of the state, this means that it is difficult to control and potentially threatening to social stability. This has a bearing on its ability to build public trust and social acceptance, on its fundraising and personnel recruitment and, eventually, on the effects of its policy advocacy (Hildebrandt, 2011; Zhan and Tang, 2013: 385). Some NGOs consciously refrain from registering, opting to work on the margins; others insist that registration is not worth the trouble, as the organization's aims are limited in scope and time. But as Hsu and Hasmath argue, an NGO's official registration can be interpreted as confirmation of the value that the government attaches to the organization's work, which can result in social or political capital that helps the NGO in its activities (2014: 529).

The registration process is complex and time consuming. An organization must first be examined by a government bureau with a connection to the NGO's area of interest. After this screening, the actual application can be made to the Ministry of Civil Affairs. The relevant government bureau then plays a supervisory and leadership role for the NGO, assuming the responsibility for its financial and political affairs. Registration must be renewed annually, with the possibility that an application for renewal is rejected (Schwartz, 2004: 37-38). Some of the problems NGOs encounter in the registration process is also reflected in the fact that the term *fei zhengfu zuzhi* ('NGO') does not have a particularly clear or consistent definition, either legally or popularly. It is regularly used interchangeably with *shehui zuzhi* ('social organization'), *gongyi zuzhi* ('public benefit organization'), *cishan zuzhi* ('charitable organization'), and *minjian zuzhi* ('popular organization'), which indeed are all domains in which NGOs can be active (Hsu, Hsu, and Hasmath, 2017: 1158-1159). In terms of staff members, NGOs tend to be quite small (Teets, 2013).

GONGOs

Over time, NGOs have become very active players, or at least try to be, in many sectors of Chinese society. Data on the number of domestic NGOs are difficult to collect and often contradictory. The frequently quoted number of 431,000 officially, government-registered NGOs in 2009 sounds plausible, but it leaves out the NGOs that find the process of registration impossible to comply with, or that have decided to continue their work unofficially. In 2008, there were 3298 ENGOs active in China (Gao, 2013; Tsang and Lee, 2013: 155, 156, 157). Estimates from 2013 arrive at a total number of active NGOs of 546,000, but do not distinguish in which fields these organizations are active (Hsu, Hsu, and Hasmath, 2017: 1158). Apart from these organizations, a number of groups that are active in the environmental field and elsewhere are better identified as government-organized NGOs, or GONGOs; alternatively, these groups are known as PONGOs, or Party-organized NGOs (Yuen, 2018). GONGOs are financially dependent on the government and 'operate in the policy domains related to the agendas of their official supervisory agencies, i.e., government mandated functions such as disease prevention and social-welfare matters' (Tsang and Lee, 2013: 156-157; Schwartz, 2004). GONGO leaders as well as staff members have sometimes served previously in official state capacities, and the contacts they built in their earlier positions can make their GONGO work easier and more successful (Schwartz, 2004). Well-known early examples of GONGOs include the All-China Women's Federation and the Chinese Youth League, which originally started out as CCP mass organizations reaching out to discrete segments of the population (Ho, 2008). As a result of decentralization, professionalization, and the withering away of the bottomless funding that supported their work, these mass organizations have been forced to more actively find relevance without state backing. Because of their close ties with the state, it sees GONGOs as 'safe', and this often makes them more successful in advocating policy changes (Teets, 2014). At other times, their state connection forces them to take on government tasks they would rather not engage in. Although they can be critical of the actions of lower administrative organizations, the central leadership and its decisions are out of bounds (Schwartz, 2004; Dai and Spires, 2017). GONGOs encounter less problems when interacting with foreign counterparts, have easier access to government officials and official data, and are able to obtain financial support from both international and domestic organizations (Ewoh and Rollins, 2011). Domestic NGOs, on the other hand, ideally stay clear of foreign financial support, as this creates suspicions in the party-state. To function, they have to avoid any suggestion that they are manipulated by foreign interests that seek to create a process of peaceful evolution (or transformation) that is harmful to both social stability and regime legitimacy (Teets, 2013).

Well-known examples of environmental GONGOs are the Centre for Environmental Education and Communications (CEEC), the Policy Research

Centre for Environment and Economy (PRCEE), and the China Environmental Protection Foundation (CEPF) (Schwartz, 2004). They all resorted under SEPA, then MEP, and now MEE. The CEPF was created in 1993 under the wings of the State Environmental Protection Agency and financed with the United Nations Environmental Protection Prize that had been awarded to Qu Geping, chair of the Environmental Protection Committee of the National People's Congress and a former administrator of the Administration. The Foundation's main goal is to facilitate the donations of funds and goods to help develop environmental protection projects (Schwartz, 2004; Ewoh and Rollins, 2011: 50; Tsang and Lee, 2017).

Embeddedness versus consultative authoritarianism

While NGOs with political missions and activities tend to run into major obstacles and meet active political obstruction, environmental NGOs that focus on a wide variety of environmental activities and garbage disposal-related initiatives are occasionally more successful, as their work complements state policies and policy goals (Hildebrandt, 2011; Wu and Chan, 2012). Friends of Nature, for example, collaborated with the Beijing municipal government in a waste classification and separation project in 2012. They entered a number of residential communities and struck camp there for an extended period of time to spread the gospel of garbage classification and separation and educate the residents (Friends of Nature, 2013). But these are exceptions, as NGOs try to stay in the background as much as possible, concerned as they are about maintaining good relations with the government and avoiding potential trouble as much as they can (Teets, 2013). As a result, they shy away from actively and openly supporting citizens' protests or actions. NGO participation in public protests and actions largely depends on the nature of the events, but also on where they take place. In some parts of the country there is a friendlier climate for ENGOs than in others; the Beijing climate is not friendly toward NGO activities (Spires, 2011). This general reticence leads to grumblings among the people that ENGOs are seemingly more interested in creating a united front with the government than siding openly with their demands, of whatever type. However, by adopting such strategies and accepting the fact that they are embedded in a broader political structure, they are able to circumvent the stringent regulations that make their activities so problematic in the eyes of the state (Ho, 2007; Teets, 2013; Yuen, 2018). In addition, green activists make avid use of informal networking opportunities with Party and state officials, as this can increase the effectiveness of their campaigns. With great hesitation, the state has even allowed some NGOs at the negotiating table. This has brought forth what Jessica Teets calls consultative authoritarianism, a system that encourages the simultaneous expansion of a fairly autonomous civil society and the development of indirect tools of state control, ultimately leading to more regime legitimacy (Teets, 2013: 2, 3). This mode of operation works especially well at the levels of the local governments. It also offers opportunities to NGO staffers to cross the divide and join the other camp to embark on a more rewarding, or at least more socially respected, career than activism.

A number of independent ENGOs are also stepping up their activities, sharing their knowledge with government agencies, offering policy suggestions, writing petitions, using media outlets and social media to draw attention to urgent cases, discussing policy alternatives with officials, and providing legal assistance to pollution victims (Yuen, 2018). None of these activities are appreciated by local governments, particularly when these ENGOs play an active watchdog role on governmental compliance with environmental legislation. Yet these same lower levels of government are also very keen to cooperate with ENGOs, particularly when they are under pressure to follow up a drive organized by the central levels. Lower administrative bodies then turn to ENGOs to help solidify practices and strengthen the implementation, control, and enforcement of existing policies and projects, as ENGOs are much more familiar with local conditions and aware of relevant activist sections of the population that can make a drive successful in practice. In short, ENGOs are occasionally invited to support government policies, but they are never asked to participate in the design and formulation of the policies, or to make the voices and opinions of the people heard during the decision-making process.

The 2015 'Opinions of the CPC Central Committee and the State Council on Further Promoting the Development of Ecological Civilization' hold out the promise that the role of public participation will be expanded, specifically referring to the participation of ENGOs in environmental governance and greater environmental transparency (Central Committee, 2015; Geall, 2015a). These 'Opinions' are the first instance where the term 'civil society' is officially mentioned, with the declaration that the government should 'actively promote the third-party treatment of environmental pollution and introduce non-governmental organizations to take part in the treatment of environmental pollution' (Geall, 2015c). It promises that ecological civilization 'will expand public participation in the initiation, implementation and postassessment of construction projects in an orderly manner' and that it will 'guide all types of social organizations [...] to pursue healthy and orderly development and give play to the role of NGOs and volunteers' (Geall, 2015c). In practice, however, none of the promises held out in the 'Opinions' have been followed up by an actual improvement of the conditions that ENGOs are forced to work under.

ENGOs in Beijing

For this project, I contacted a number of domestic ENGOs in Beijing that had been very active in raising the consciousness of the population concerning garbage disposal, garbage separation, and the benefits of these activities for improving the living environment. I had the pleasure to meet with representatives of Friends of Nature China, the Green Beagle/Darwin Institute, Huan You Science and Technology, and Hong Chao, who were willing to exchange their views and opinions with me. Hong Chao is not an ENGO in the strict sense of the word, but rather an environmental entrepreneurial player; their Red Nest project was discussed in more detail in Chapter 3. Following Tsang and Lee's typology, Hong Chao can be seen as representing that part of the middle class that is making use of ENGO activities to ride the wave of environmentalism. It is more focused on expanding its relations with government officials and companies to support its own business interests than on advocating for any specific environmental interests (Tsang and Lee, 2013: 158).

Aside from wanting to find out more about the activities these organizations had been engaged in, or were planning to start, I was interested in particular about their impressions and evaluations of the app-using O2O recycling companies like Incom, Taoqibao, and Zaishenghuo that were introduced in Chapter 2. I wanted to know whether the ENGOs thought these initiatives offered new and relevant ways to abate the waste crisis; whether they contemplated cooperating with them or even support them; whether they thought these companies would be able to improve the classification and separation of waste; and whether they had an opinion on whether these companies could play a role in improving the lives and working conditions of the waste pickers they employed. Moreover, I was interested in their attitudes towards the incineration projects contemplated by the government, and their impression of and potential involvement in popular actions organized against these projects.

Much to my disappointment, I discovered that while all of these ENGOs had initially focused on waste and garbage disposal and the wellbeing of those working in the business, their attention had now shifted to other campaign objectives. Some of them had reacted to the increased concerns about air pollution and 'blue days', particularly PM2.5 (particulate matter) pollution, as voiced by both urban residents as well as the local and national governments (Wang, 2016; Kennedy and Chen, 2018). In light of the blue sky push, coupled with explicit utterances by national and municipal leaders that PM2.5 pollution needed to be dealt with as soon as possible – up to the point where local administrators were threatened by dismissal when they were not able to improve local air quality – the ENGOs were aware that this was a topic that was not contested and there would not be problems in the activities they planned related to it (Kostka, 2014; Kennedy and Chen, 2018). Some ENGOs explained that while the waste problem continued to be urgent and needed to be solved, and while they continued to be worried about the fate of the waste pickers, it was no longer the main concern of their organization. Friends of Nature, for example, was still engaged in various drives in Beijing aimed at education and raising popular awareness about the need to separate garbage and produce less waste. This was translated into various zero-waste activities they organized in individual residential communities. On a national level, however, they had become active in broader environmental campaigns, such as an effort to save the Yunnan peacock, at the time I interviewed one of their staff members (Interview Friends of Nature, 2017).

ENGOs, waste, and O2O-companies

All the ENGOs I talked with were more or less familiar with the emergence of O2O initiatives in Beijing, but none of them had had seen them in action.

Green Beagle

Green Beagle had heard of the smartphone application developed by the Tao-qibao company but had never seen it in actual practice and was not familiar with how it worked. Ms. He, the staff member, agreed that a mobile service looked like a good idea and might efficiently solve some of the problems that existed in the waste disposal situation. But Green Beagle was certain that the problem could not be solved in its totality by merely depending on apps; there was simply too much garbage that was dumped, and the demands to solve the issue were too complex and too many. Moreover, the size of the O2O companies suggested that they were operating on a level that was too low and too limited. Their activities might bring some alleviation of the problem, but their scope was too small to make a big difference.

As for the contribution of smartphone apps to the improvement of the wellbeing of the people working in the waste picking business, earlier Green Beagle research had uncovered that these workers are generally exposed to the dangers of secondary pollution while they do their work. The O2O companies might have the opportunity to offer better worker protection, but this depended on the exact contents of the contracts they offered to their employees.

In terms of educating the people, Green Beagle was convinced that propaganda and public service advertising would make a difference in changing the minds and actions of the people, but that it would be a long process to engineer concrete changes. Ms. He suggested that adopting formal rules and regulations would be more effective. In other words, government actions would lead to more results than other initiatives. These government efforts might include rules that would lead to a decrease in garbage production; levying individual charges on the amount of garbage that is dumped, i.e., on the basis of the principle that the user pays; levying a charge on the use of plastic bags (on the basis of experiences gained in Taiwan and Japan), etc. The government could contemplate such initiatives, but they would only be effective as long as the people clearly knew what and what not to do.

Huan You Science and Technology

According to Mr. Zhang, who represented Huan You, the smartphone applications were inspired by big data operations such as TaoBao, the large and popular online shopping entity run by Alibaba. Because the use of the app is so similar to online shopping, Huan You was convinced that it would appeal more to the younger generations through its familiarity and convenience. Mr. Zhang was not convinced that the O2O companies and their apps would help decrease the amount of garbage, as he was confident of the public's ability to recycle, particularly when there was money to be made. In Huan You's view, the success of these apps ultimately depended on whether they would be able play a role in disposing of the garbage that was difficult to recycle and had no monetary value.

Regarding the improvement of the lives of the waste pickers who were absorbed into O2O companies, Huan You was cautiously optimistic. Ideally, the same people who were active as informal waste pickers would become O2O workers; in the best scenario, their work style would be updated, and their service levels would improve. This would have a positive influence on the population's willingness to turn to them. When it came to recycling and garbage disposal in general, however, Mr. Zhang sounded rather pessimistic. Residents might recycle and sell some valuable junk, but the rest would just be thrown away. Waste collectors might take some of the valuable recyclables out of the waste stream, but not all of it; they were certainly not as efficient as people tended to make out. As a result, the major part of the waste still ended up in landfills or incinerators.

In general, Huan You rued the lack of knowledge and awareness of the people concerning garbage production and reduction, although Mr. Zhang had to admit that the situation in the larger urban areas, i.e., the tier-one and tier-two cities, was much better than in the countryside. He recounted how in the past his organization had once converted the gutter oil they had collected into soap bars, to provide an educational experience in recycling. The public, unfortunately, thought this type of soap could not be hygienic and did not want to buy any of it. In the end, they decided to use the gutter oil to make fertilizer and enzymes instead. His organization was not actively involved in educational activities at the time of the interview and had no plans to so in the future.

Friends of Nature (FoN)

Ms. Lin, the representative I interviewed, was the most outspoken and expressive of the ENGO members I spoke with. FoN was cautiously positive about the O2O apps that were in use because they offered residents new ways and methods to deal with waste. When these companies were part of or owned a larger system of recycling and disposal, including large recycling facilities, this could lead to less secondary pollution overall. Moreover, in the experience of FoN, some of these companies actively cooperate with communities by absorbing the original informal garbage collectors into their ranks and making the market more professional on the whole. However, funding was needed for people to pick up recyclables at individual addresses; the companies either needed to explore alternative sources of financial support or turn to the government to subsidize them. FoN was also convinced that O2O companies were only interested in valuable recyclables like used smartphones. Finally, the organization expressed doubts about the effects of the educational materials that these companies sent to their application users.

FoN had actually been courted by some of the O2O companies, although not the ones that were looked at in this project. O2O companies are interested in collaborating with ENGOs because they have excellent lines of communication with university authorities, departments, and students. On a side note, students are drawn to ENGOs like magnets. Many are attracted by the public visibility offered by the media. Most of the environmental activities also are seen as meaningful and fun experiences for self-exploration and socializing, including training in leadership, skills in interpersonal relations, and exposure to new horizons of life (Yang, 2005: 62). This base of young collaborators offers the companies opportunities for advertising as well as potential markets. Moreover, the companies are interested in making use of the volunteer resources of the ENGOs for their own promotions and activities. And, although they would probably never admit it, they need FoN's practical experience and contacts in setting up recycling drives within the residential communities.

In the field of education about separation and recycling, FoN has continued to be very active, organizing various online and offline activities to stimulate public awareness. They have come to the conclusion that merely putting up posters and distributing booklets and leaflets is not enough; people need to be taught in practice about all the details of recycling. FoN has developed many cooperative relationships with educational organizations and residential districts. As a result of these connections, they can offer lecture series and teaching programmes for elementary and high school audiences, as well as for a more mature public; cooperate with college clubs; and organize events where schools and families interact. Similar to the philosophy embraced by Ms. Zhang of the Solid Waste Management Office introduced in Chapter 5, FoN is committed to encouraging school children to influence their parents with the knowledge of garbage recycling that they have picked up in school. During recycling drives they organized in residential communities, FoN has given small financial rewards or gifts to residents who disposed of their recycled garbage at designated spots, or who have classified their garbage into bags with RFID-codes attached. During other local events, like sport tournaments, FoN has promoted awareness of the need to use fewer single-use glass and plastic containers. FoN also organizes tours to landfills and incinerator facilities, usually followed by feedback sessions (see FoN member Lianpeng's review of a visit to the Gao'antun incinerator in Chapter 7).

FoN also proactively submits plans and proposals to local and national government bodies, hoping that they will be adopted and be discussed. On the whole, their track record has not been very successful in this respect. For the 13th Five Year Plan adopted in 2015, FoN submitted various proposals and amendments in collaboration with other organizations (i.e., the State Environmental Protection Volunteers Association from Wuhu City, the

Shenzhen Zero Waste Environmental Protection Public Welfare Development Centre, and others) (Friends of Nature, 2017a, 2017b). These focused, among other issues, on the need to pay more attention to the front-end generation of garbage rather than the back-end incineration, claiming that the planned steps would mean a retreat from the preceding 12th Plan, effectively negating any positive outcomes generated during the preceding planning period. None of the ENGO's suggestions were adopted or referred to in the final version of the plan. One can debate whether continuing to provide these suggestions, or offer advice to government bodies in general, is worthwhile, as none of it ever seems to be picked up. On the other hand, showing that an ENGO is actively designing alternatives, is thinking with rather than against the government, certainly contributes to creating an impression of support. Publicizing these advisory efforts also demonstrates the relevance of the ENGO to its support base as well as other interested parties.

Hong Chao

Hong Chao did not have an opinion on the activities of the O2O companies. It had designed a smartphone application itself and was in the process of rolling out a garbage sorting system of its own, thus turning it into a competitor. Hong Chao stressed that in the Red Nest model that it had developed, waste picker welfare was considerably better than elsewhere and would leave the working conditions of O2O employees far behind (see Chapter 3).

ENGOs and Beijing incinerators

Green Beagle

Green Beagle is convinced that the government prefers waste incineration over landfilling because of the limited space available and exorbitant land prices. Opting for landfills has many attendant problems related to the costs of the protective measures that have to be taken. The fact that incineration produces electricity also appeals to the government, although the Chinese diet produces wetter garbage, leading to a lower calorific value when it is incinerated. This problem could be offset by adding a combustion improver (kerosene, coal) or by introducing an extra step in the process, such as dehydrating the wet garbage before burning it. Both options make the process costlier, and adding a combustion improver leads to more potentially toxic air pollution. As to the fierce NIMBY protests in Beijing and other urban areas against incineration that have been reported in the Chinese (social) media as well as in the West, Green Beagle was convinced that these events did not necessarily mean that the public prefers the use of landfills over incinerators. Rather, the people are opposed to both solutions, particularly when they live close to such projects. Green Beagle asserted that sometimes public opinion is able to delay or change municipal plans for incineration or landfilling, referring to the postponing of the plans for the Liulitun and Dagongcun incinerators, but criticized these protests because they were motivated only by the public's self-interest and not by fundamental environmental concerns. In the end, these protests are unreasonable because when incinerator factories are built in other, more distant locations, the residents will have to pay a higher sanitation fee. This in turn will also lead to popular dissatisfaction. On the whole, Green Beagle was not a supporter of NIMBY activism and did not participate in it.

Concerning the incineration technology that was presently in use, Green Beagle pointed out that the biggest problem was the public's lack of trust in the government and the way it operates the facilities. Although the technology may be imported from abroad, may be state-of-the art, completely safe, and perform in agreement with international requirements, in practice all of the garbage recycling procedures remain too secretive and the control mechanisms contain too many loopholes. As a result, incidents and accidents continue to occur, and this leaves the public with a bad impression.

Huan You Science and Technology

Huan You is convinced that the Chinese government lacks a clear understanding of garbage, especially when compared to Western nations. The government has a preference for garbage incineration because it is convinced that it means they will not have to come up with or invest in methods of recycling. When applying the recycling standards that are in force in developed countries, 90 percent of the garbage produced in China should be recyclable. Mr. Zhang was certain that although incineration technology is well developed, in practice it does not perform well, and certainly not well enough to live up to the commitments that China has agreed to at the International Climate Change conferences. There is insufficient control and oversight in the incinerator facilities. Mixing large amounts of wet garbage and glass prior to incineration produces dioxin (TCDD). It would be easy to prevent or reduce the production of TCDD by separating the garbage more carefully. Moreover, the incinerator factories are established by companies that have personal connections with the decision makers in government circles; this makes it a market where ordinary (i.e., unconnected) companies cannot compete freely on the basis of their qualities. This can lead to a situation where suboptimal solutions are chosen.

As for the potential effects of NIMBY protests and other popular actions against incineration, Mr. Zhang was not very hopeful. Like Ms. He at Green Beagle, he deplored the lack of fundamental awareness among the participants and lamented the fact that the government was all-powerful in pushing through its plans. Huan You did not actively support NIMBY activism.

Friends of Nature

The organization is worried about the speed of the development of incineration because, in its opinion, the front-end part of the recycling process is not completed, resulting in large amounts of garbage that are burnt in the back-end, leading to major problems of secondary pollution. The incinerators are funded by the government and solve the garbage problems that the government is supposed to deal with. It is hard to change this situation. Ms. Lin was convinced that the technology currently in use can burn much more garbage than is actually produced. And although incinerators are designed to decrease the amount of garbage that is generated, they in fact demand that more garbage be produced in order to be more profitable. As for the way the incinerator facilities operate, FoN finds it hard to ascertain whether they function as they say they do. It is extremely difficult to gather data, both at the level of the government and of the incinerators themselves. Monitoring systems, whether online or otherwise, allow room for fraud. Factory inspections, if they happen at all, only take place twice a year, when the external conditions are at their best and the lowest levels of secondary pollution are recorded.

As for popular protests against the construction of incinerators or landfills, FoN is convinced that the government's overriding concern with maintaining social stability make any real action impossible. Ms. Lin explained that her organization does not dare to support public protests in fear of the political fallout. In the current climate that puts a premium on maintaining stability, organizing the public on a large scale to pursue environmental issues is impossible. Thomas Johnson wrote that although FoN explicitly did not support the Liulitun anti-incinerator protests in 2007, it did organize garbage classification and separation projects in the involved communities (Johnson, 2013b: 370). This manner and level of collaboration corresponds with the position FoN took in 2017.

However, FoN pointed out that in some more recent protest movements, the participants had announced that they were not opposing the construction of an incinerator because they did not want it in their neighbourhoods. Instead, they expressed a desire to supervise the work, so that it would operate in a more environmentally friendly and efficient way. If both the government and the ENGOs could lead protest movements toward such a direction, they would be more useful.

Hong Chao

Hong Chao, represented by Mr. Cui, was very supportive of the trend towards incineration. According to him, the technology is fully developed. However, the public does not know enough about the processes that take place in the incinerator. Hong Chao was convinced that the public will accept incineration once it understands how it works. Having said that, Mr. Cui suggested some improvements in the communication pattern between the public on the one hand and the government and the incinerator companies on the other. The government should listen more to the opinions of the people living near the sites selected for construction. And the companies operating the facilities should be open and above-aboard about the exact amounts and effects of the dioxins and smell that are released during the incineration process. Keeping the people in the dark about these aspects results in unnecessary panic and resistance.

In support of popular actions

Despite the hesitancy about supporting popular actions that I encountered among the ENGOs, they were not fundamentally opposed to them. Instead, they expressed their doubts about whether such actions would lead to any satisfactory results. After all, these organizations have been moderately successful in addressing concerns and grievances among the people on the one hand and the government on the other. Their embeddedness, however, has made it difficult for them to act openly (Van Rooij, 2010; Yew, 2017). Various studies have analysed how ENGOs, by operating in the background, have been able to defend the rights of inhabitants in local cases. These include the provision of legal advice and support in a number of environmental information disclosure requests related to air quality and pollution in which FoN and others were involved (Schwartz, 2004; Wang, 2016), support for the growing 'no burn' constituency that critically follows government plans to construct new incinerators (Bondes and Johnson, 2017; Interview FoN, 2017), and legal assistance for pollution victims (Van Rooij, 2010). Lawful activism, i.e., suing local authorities in court, has been the mode of operation increasingly chosen by ENGOs. These processes are helpful in instilling legal awareness among participating residents, and show avenues for action that do not necessarily jeopardize the people who take part and still lead to results (Yew, 2017).

In all of these cases, the links between ENGOs and the (Chinese) media have played an important role in whether popular action of any kind will have an effect. When the media becomes aware of an event and decides to publicize a case, it can resonate with similar occurrences elsewhere in the country. This is a development that the party-state fears greatly, as it could galvanize larger groups of people into action – and this by itself is a threat to social stability. Media reporting also helps bring grievances to the attention of the government, which may (or may not) decide to step in. Moreover, ENGOs have been very successful in making use of the opportunities that social media such as Weibo and WeChat offer, particularly when it comes to mobilizing support among larger groups of likeminded people. This has been possible despite the fact that they have to negotiate endless suppression and censorship. Yet even positive and helpful media attention has a negative effect, as it can lead to government interference with NGO activities. An NGO's premises may be searched or closed down; its staff members may be subjected to harassment or invited to 'have a cup of tea' at the police station – a euphemism for interrogation. In the worst case, NGO staff members can be prosecuted when an NGO's activity is seen as overstepping the bounds (Spires, 2011; Teets, 2014).

7   The Politics of Incineration

The Chinese government sees the incineration of garbage as the most effective solution for the problem of disposing of the ever-increasing amounts of waste generated by the steadily expanding populations in the urban centres (Yang, 2013). Good governance is generally considered a prerequisite of urban health. Indicators for the level of modernity, development, and social progress of a state include setting up health governance organizations and institutions, such as the provision of services for sanitation and garbage removal and disposal (Kickbusch, 2007; Vlahov et al., 2007). Beijing and other Chinese metropolises all aspire to be seen as proper modern (global world) cities, using Singapore as their inspiration. All of these cities have top-down organized garbage disposal structures in place, with garbage trucks, transfer stations, landfills, and incinerators. Informal garbage collecting/picking may be more efficient than automated processes (and cheaper, since it contributes raw materials back into the production process), but it is undertaken by an unregulated 'population [which] can be theorized as a kind of disposability and throwing away within capitalism' (Yates, 2011: 1680). Added to this is that informal waste picking is less efficient in practice than we often assume or are led to believe. All this is seen as a blight on the modern, well-organized image of the Chinese city that the government wants to project, and this reflects badly on the officialdom that manages it (Ou, 2011; Williams, 2014: 196). But how many incinerators will China need to cope with its garbage?

The 13th Five Year Plan (2015-2020) states that in economically more developed areas and cities with land shortages and a large population, incineration should be designated as the priority technology for waste disposal, while the construction of such plants in third- and fourth-tier cities must increase (Li et al., 2015; Nelles et al., 2017). According to official estimates based on the amount of waste generated in 2011, some 600-700 facilities would need to be built, each with a capacity of 1000 tons per day (Yang, 2013: 182). The China National Renewable Energy Centre 2017 report estimated that 277 million tons of collected MSW will need to be disposed of by 2020, reaching 369 million tons by 2030 and 409 million tons by 2050; 87-90 percent of this amount will be available for incineration and heat generation (Energy Research Institute of Academy of Macroeconomic Research and National Development and Reform Commission, 2017: 316). On the other hand, the people – and not only those living near incinerator factories – are not convinced by these arguments. According to Yang

Changjiang, quoting poll results from 2009, '92% of people believe that waste incineration will harm human health, and 97% are against incinerator construction' (2011: 190), despite the authorities' assurances that the processes within the facilities are up to national and even international standards (Yang, 2013; Interviews, 2017).

Landfills

While landfill disposal used to account for 80 percent of the waste being treated nationwide, this option is increasingly seen as undesirable and too costly. As a result, the percentage of landfill disposal had dropped to 63 percent by 2013 (Li et al., 2015: 234). Landfills do have the potential to create large numbers of jobs for informal waste pickers who reduce the amount of waste at the tipping point, but this option is rarely seen as a reason to continue operating them (Hoornweg, Lam, and Chaudhry, 2005; Dorn, Flamme, and Nelles, 2012). More than half of the landfills still in operation have seen serious surface water and groundwater contamination due to the lack of leachate collection and treatment systems; many have also collapsed, leading to serious pollution issues (Yang, 2011). The presence of waste pickers at landfill sites is seen as a source of problems related to hygiene and control. Putting new sites into operation with the required protective measures against secondary pollution will require prohibitive investments (Hoornweg, Lam, and Chaudhry, 2005). On a more structural level, opening landfill sites calls for an amount of land that is simply no longer available in the vicinity of Chinese cities (Cheng and Hu, 2010; Yang, 2011; Yuan and Li, 2017). This has not deterred some people, mostly residents, from suggesting that the garbage could be moved out of the densely populated areas and be buried or landfilled elsewhere, such as in Inner Mongolia or in the desert of Xinjiang Province (Johnson, 2013a; Interviews, 2017). Some urbanites even joked that urban garbage could be exported to Africa (Interviews, 2017).

Around Beijing alone, more than 330,000 square meters is taken up by landfills, a figure that does not include the myriad illegal landfills that are also in operation (Watts, 2010; Wang, 2011b). Increasingly, urban residents are resisting the opening of new landfill sites, expressing worries about the potential of pollution and more importantly the foul odours and presence of waste pickers (Xie, 2011; Dorn, Flamme, and Nelles, 2012). To counter these latter complaints, some past municipal administrations have resorted to using deodorant guns to mask the stench, but this can only serve as a temporary solution (Watts, 2010). In 2009, the Beijing municipal government expressed its intention to eventually phase out the use of landfills completely (Yang, 2011: 192).

The presence of 500 illegal and semi-legal landfills surrounding Beijing, most of them open dumps, has been documented by the photographer and journalist Wang Jiuliang, who was the first to bring their environmental and human impact to the attention of officials and residents in Beijing and China at large. Many interested parties elsewhere have also reacted strongly to his work. In 2007, Wang started following the waste collectors and those involved in moving collected junk out of the city, photographing their activities and those of the people working and living on and near the dump sites. He made visible how migrant children played in the dirt and with the garbage, including using discarded hypodermic needles and other harmful waste as toys. Wang showed how sheep and goats were feeding on the garbage, thus potentially bringing contamination into the food chain. He gave an impression of the squalor of the landfill sites and the people existing there. By using the sites' GPS coordinates and locating them on Google Maps, a project that he started in 2008, Wang demonstrated that Beijing is surrounded by a so-called Seventh Ring Road of rubble sites and *lajicun* ('junk villages'), 500 in all (Kao, 2011; Wang, 2011a; Wang, 2017; Kao and Lin, 2018). The photographs were exhibited in the Beijing Songzhuang Art Museum in 2010 (Kao and Lin, 2018: 301). Wang used his photography as a starting point for the production of a documentary called *Beijing Besieged by Waste* (垃圾围城, also known as *The City Besieged by Waste*), which was released in 2011 (Wang, 2011b). Wang's virtual Ring Road of Junk was actually located between the Fifth and Sixth Ring Roads at the time, but has been forced further out into Beijing's suburban districts over the intervening years. The impact of Wang's documentary was enormous, both domestically and internationally. Some sectors of the Beijing municipal government assisted Wang and supported his work to strengthen their own position, providing him with access to the media and connecting him to other relevant contacts; other sections of the bureaucracy even seized upon the term 'besieged by waste' to advocate the increase and faster development of incinerator facilities (Kao and Lin, 2018: 301-303). Wang's documentary succeeded in making the highest levels of the government aware of the seriousness of the situation, including former Prime Minister Wen Jiabao. Since the first screenings of the documentary, steps have been taken to clean up (some of) the mess and to organize and regulate garbage collection and treatment (S. Lu, 2017). As described in Chapter 4, by late 2017 the municipal government had started the comprehensive clearing out of junk villages and their inhabitants.

Biological treatment

The biological treatment of waste also takes place in Beijing on a very limited scale at Beijing Municipal Chaoyang Circular Economy Industry Park at Gao'antun. This treatment method only works well when the waste is properly sorted, for example when it is used to treat kitchen waste. When it fully uses the organic materials in waste, it causes less secondary pollution and is easy to control (Li et al., 2015). However, composting takes place over a long period of time and needs large land resources, making its end product expensive to sell. Another negative aspect is that there is currently no great demand for reusing waste that has been composted or gone through a process of fermentation. The fertilizer generated by biological treatment has low nutrient content and contains certain heavy metals, so it can only be used as a soil modifier and cannot replace chemical fertilizer altogether (Xu, He, and Luo, 2016: 38). This has not made peasants eager to apply this organic fertilizer in lieu of the inorganic variety, even though they are aware that the latter is harmful to the soil (Wei et al., 2000). The combined result of these factors is that the number of biological treatment facilities nationwide, as well as their treatment efficiency, has seen a steady decline since 2004 (Nelles et al., 2017). If a scheme for the separation of wet waste or kitchen waste at the level of the residents were designed, this type of garbage, properly sorted, would be eminently suitable for biological treatment.

Incineration

Of the three options for waste treatment, incineration is seen as the most promising because it is said to have nothing but positive benefits. Incineration solves the ever more pressing problem of burgeoning waste in one fell sweep: by burning it, the waste simply disappears. Moreover, burning garbage produces energy (WtE) that can be used for a variety of purposes. On a more symbolic level, the embrace of incineration technology demonstrates to the world how modern, developed, and evolved China is. Over the years, incinerator technology and/or equipment has been imported from abroad. Companies from Japan (Mitsubishi, Hitachi, and others), Germany (Stan Miller), France (Alstom), Switzerland, and other countries have transferred technology and/or equipment. Chinese companies such as HuaGuang Shares, Huaxi Energy, and Chongqing Three Peak Covanta have digested foreign technology and combined it with independent research, demonstrating the technological development of the country. Increasingly, Chinese domestic companies design and build incinerators; these include the Everbright Environment Co. Ltd. from Shenzhen and the New Century Energy and Environmental Protection Co. from Hangzhou (Chin, 2011; Li et al., 2015). With the focus on incinerator construction, less attention is paid to the negative side effects of the processes that take place within the plants, but there are important questions that need a satisfactory answer. How can the secondary pollution caused by incineration be avoided, or at least limited as much as possible? How can the rest products from incineration such as fly ash, flue gas, and other toxic substances be disposed of as safely and least intrusively as possible (Cheng and Hu, 2010)? The 13th Five Year Plan currently in force calls for the consideration of these issues, but fails to provide specific measures (Nelles et al., 2017). It is too early to tell whether the 'Action Plan for Straightening out the Municipal Solid Waste Incineration Power Generators to Meet Emission Standards', promulgated in 2018 by the newly reorganized MEE, will have a positive effect (MEE, 2018).

'Wet' MSW

The available literature on MSW in China, including the waste produced in Beijing, agrees that it has special characteristics. It is made up of food waste, paper, textile, rubber, plastic, glass, metals, wood, garden waste from public parks and green areas, street cleaning waste, and miscellaneous types of inorganic waste (e.g., stones, ceramics, rubble from construction sites, and ashes) (Dorn, Flamme, and Nelles, 2012; Linzner and Salhofer, 2014). As I have shown in preceding chapters, the official municipal waste collecting authority does not engage in any systematic sorting and recycling of this waste; instead, it only tries to rake in as much garbage as possible to increase the amount of government subsidies it can claim (Yang, 2013: 177). Large numbers of informal recyclers sort and recycle a large volume of valuable materials with resale value, including paper and plastics, out of the waste stream. However, according to some of my sources, the importance of the recycling activities of this informal sector should not be overestimated. According to them, this could be done much more systematically and on a much larger scale (Interview with Hong Chao, 2017). Metal waste remains under the monopoly of the government's recycling facilities and their collaborators; informal waste sorters are legally excluded from dealing in it, although they cannot resist the temptation of picking up and selling metals of any kind whenever they encounter them (Minter, 2013a, 2013b, 2015).

The most distinctive characteristic of Chinese MSW, however, is that food waste makes up the largest part of it, from 70-80 percent. This causes Chinese MSW to have much higher moisture levels than that in Western countries, i.e., 78 percent versus 12 percent (Cheng and Hu, 2010: 3819; Dorn, Flamme, and Nelles, 2012; Interviews, 2017). When looking for reasons why this was the case, I heard some interesting explanations. One of the experts I interviewed, an academic, was convinced that people 'in the West' only eat hamburgers packed in little boxes, with the result that, in his opinion, Western MSW mainly consisted of paper and cardboard. Chinese waste, on the other hand, was full of bones, vegetable peels, watermelon rinds, and so on (Interview with Hong Chao, 2017). Taking his argument to its logical conclusion, the wetness of Chinese MSW was the direct result of the Chinese diet, and was rooted in culture.

Because of its wetness, the calorific values of Chinese MSW are less than half those in more developed countries (Cheng and Hu, 2010; Xu, He, and Luo, 2016). This creates a range of problems at the moment of incineration, including difficulties in ignition, an unsteady and unstable combustion flame, incomplete combustion of the waste, and the increased formation of air pollutants. Lower combustion temperatures increase the dioxin levels in the flue gas (Nowling, 2016). To incinerate, pre-treatment of the MSW is needed. Various methods can be used to pre-dry wet waste (Interviews, 2017). In many cases, supplementary but higher-cost fuel is added, which decreases the net gain in energy and also increases the operating costs of the facilities (Cheng and Hu, 2010: 3819). Adding feedstock like coal is allowed, but the amounts that are added reportedly exceed the legally permitted limits; the stipulated percentage is 20, while cases of up to 70 percent feedstock have been documented (Balkan, 2012). These additions cause wear-and-tear to the equipment, calling for more frequent replacements than initially calculated. The plants' emissions also contain more polluting elements. Another reason to add feedstock to waste has nothing to do with its wetness: some individual incinerator operators add large amounts of coal to the MSW mixture to achieve a greater power generation capacity. This entitles them to more electricity subsidies from the government. This practice not only consumes more resources instead of saving them, but it also generates more pollution – contradicting the arguments in favour of adopting incineration as a solution and harming the spirit and intention of the process of sustainable development that the laws and regulations attempt to create (Chen, Geng, and Fujita, 2009: 38).

These problems could be solved with the expansion of the separation of wet or kitchen waste at the consumer level. This wet waste, properly sorted, could be composted or set aside for biological treatment. At the same time, by making the waste that is incinerated drier without the need for costly intermediate pre-treatment steps, this would ensure cleaner emissions from the incinerators. Better sorting and recycling practices of residents would therefore also improve incineration effectiveness and contribute to a decrease in dioxin emissions (Wan, Chen, and Craig, 2015).

Liulitun and other incineration sites in Beijing

Under the 11th Five Year Plan, spanning the period 2006-2010, four sites in Beijing were designated for the construction of incineration facilities: Asuwei to the north, Gao'antun in the east, Liulitun in the west, and Nangong in the south (Xie, 2011; Johnson, 2013a). The construction plans for the Asuwei and Liulitun incinerators met fierce resistance from local residents, leading to delays or even postponement (Watts, 2010; Johnson, 2013a, 2013b). The Liulitun incinerator was planned to replace a landfill that had led to many complaints over the years because of the stench it generated. Residents living in the vicinity, many from the circles of government officials and entrepreneurs as well as the owners of apartments in high-end gated communities, rose up in protest. They were worried that the value of their real-estate investments would decline, and also expressed concerns about their health and wellbeing. Through various means, including petitioning the authorities and using their connections with officials and politicians, the residents were able to postpone the construction of the Liulitun plant. In a later phase, various considerations, including about the city's water supply (the plant would be located near the Beijing-Miyun drinking water diversion canal that supplies the city), eventually lead to the decision to drop the plan altogether in 2012 (Li, Liu, and Li, 2012: 69). Yet my interviews with Haidian sanitation workers in 2017 showed that the Liulitun facility actually was in full operation. Johnson (2013a: 123) states that once incineration had ceased at Liulitun and moved to Suijiatuo, the former facility would still remain in use for the pre-sorting of garbage.

The facility at Gao'antun, which was planned on the site of what used to be Beijing's largest landfill in the Beijing Municipal Chaoyang Circular Economy Industry Park, met comparable popular opposition, but was completed and went into operation in 2008 (Kao and Lin, 2018). It has turned a landfill, the stench of which had caused increasingly intense resistance from the people living in the neighbourhood, into a national model of modernity. The Gao'antun facility is very careful not to use the term 'incinerator' in its

Illustration 7.1   Signpost showing directions to the various facilities at the Beijing Municipal Chaoyang Circular Economy Industry Park

Author's photograph, 13 March 2017

official name. It serves as a model for other incinerator initiatives elsewhere in the country, as the regular reports about official visits on its WeChat site demonstrate (Johnson, 2013a). Images from its high-tech operations are frequently used in foreign and Chinese media reports as a template for incinerators (for example PressTV [Iran], 2014; Beijing Municipal Statistics Bureau, 2017). Beyond incineration, the plant also operates a wastewater treatment facility, medical waste disposal factory, and (since 2016) a recharging and switching service centre for electric sanitation vehicles, passenger cars, and buses in eastern Beijing (State Grid, 2016; Fieldnotes, 2017) (Illustration 7.1).

Trust

While the Chinese people are generally firmly convinced about the effectiveness of incineration and the positive aspects that incinerator technology brings, particularly when that technology is imported from abroad, there also exist deep-seated feelings of distrust towards the process as a whole and towards those responsible for it. For local residents, the potential risks that (WtE) incinerators pose to their health and living environment are the core areas of distrust (Liu et al., 2018). This was also brought forward by my sources in the ENGO community, as reported in Chapter 6. The government's credibility and people's trust in it have taken a deep dive as the result of the failure of governments at all levels to cope with increases in cases of fraud and corruption, fake products, improper academic behaviour, illegal land transactions and industrial pollution scandals, and badly functioning healthcare and civil services (He, Mol, and Lu, 2012). Without trust, it is extremely difficult to establish the cooperation and reciprocity that is needed to create support for and acceptance of incineration (Tu et al., 2011; Liu et al., 2018). As the late sociologist Fei Xiaotong argued, trust within one's kinship group or network may engender cooperation and reciprocity for a small group, but when the objects of reciprocity and cooperation are extended to include public goods and provisions, a more generalized type of trust is needed (Tu et al., 2011). This general trust is lacking in China, particularly where health and the environment are concerned (Liu et al., 2018).

Fukuyama (1995) discovered that in communities that are strongly influenced by Confucian culture, including China and other parts of the Sinophone world, there is a lack of general (non-kinship or generalized) trust. Fei Xiaotong already established that rural (or traditional) Chinese society is strongly based on kinship relations and networks, which indicates that trust is mainly influenced by high levels of kinship trust (in Tu et al., 2011; Gow, 2017); this works to the detriment of general trust. As a result of the pervasive influence of Confucian thought in Chinese society, an individual defines him/herself in terms of his/her relationships with others while adhering to culturally bound codes of etiquette, protocol, and convention. These social relations are not based on a process of voluntary association, but instead on a moral obligation (Gow, 2017). The traditional networks based on kinship relations were transformed after the establishment of the People's Republic in 1949, with the introduction of public ownership, central production plans, and centrally controlled wealth distribution. Society was reorganized in (rural) people's communes and (urban) work units, replacing interpersonal relations with a corporate framework (Tang, 2004: 5-6). With the arrival of economic reforms and marketization in the 1980s and the substitution of communist egalitarianism with phenomenal economic inequality, interpersonal trust was again strengthened (Hu, 2015).

Following Wenfang Tang's analysis, it is possible to broadly identify three social groupings that influence trust. First, relations with family members and relatives are based on kinship ties; these can be classified as parochial trust. Second, friends, neighbours, schoolmates, rural residents, people from the same geographic area, co-workers, and supervisors are part of one's immediate social and economic environment. Relations among these groupings are based on corporate trust. Third, urbanites, businessmen, out-of-towners, foreigners, and strangers are twice removed from one's immediate self. Relations among them are based on the most abstract type of trust, which can be defined as civic trust. Urbanites are part of the complex of civic trust, since they are representatives of the modern lifestyle that is marked by less corporatist community solidarity and more technologically defined civic communities (Tang, 2004: 8-9). These include the residential communities where much of the research for this present volume took place.

There is considerable academic disagreement about whether China is a low-trust society. Wenfang Tang argues that statistical evidence does not support the conclusion that Chinese do not trust each other, although ample anecdotal evidence suggests otherwise (Fieldnotes 2015, 2017; Tang, 2018). The results of many scholars also point to the conclusion that trust in institutions and political support tend to be higher in China than in other nations (Yang and Tang, 2010; Tan and Tambyah, 2011; Steinhardt, 2012; Sun and Wang, 2012; Zhong, 2014; Saich, 2016; Zhong and Hwang, 2016; Tang, 2018). There is a profound belief among the people in the benevolence of the central government and the willingness of the paramount leader(s) to right any existing wrongs. At the same time, there is broad dissatisfaction with the performance of the lower administrative levels, as demonstrated by continuing (bureaucratic) corruption and nepotism, crony capitalism, uncertainty about property rights, insufficient and/or inaccessible health-care, food safety, personal security, and police brutality (Sun and Wang, 2012; Zhong, 2014; Zhong and Hwang, 2016). In any given conflict one comes across in Chinese society, one can hear the anguished complaints that if only the (central) government knew about the situation, it would step in and take the necessary measures to end it (Zhong, 2014; Zhong and Hwang, 2016; Fieldnotes 2015, 2017). As a result, it is the lower levels of government (local governments, *chengguan* ('city management officials'), (armed) police, courts, etc.) and the institutions rendering social services to the population (hospitals, petition departments, etc.) that are identified as engaging in illegal activities and exploiting the people (Saich, 2016). For an ordinary member of the urban population, this means that every official or person in any position of authority cannot be trusted as a rule (Geall, 2015a). In 2011, the State Council recognized this by identifying the lack of credibility existing in society and lack of public trust as the main factors that have a negative influence on the wider governmental aims of progress, but little headway has been made in improving this situation (He et al., 2013).

It is this attitude of distrust that extends to the managers of recycling factories and to the officials responsible for or working in incinerator facilities. The antagonistic feeling is caused partly because nobody in the general public is really aware of what actually happens in these facilities. As a result, rumours and hearsay circulate about official incompetence and malfeasance and the real or alleged malfunctioning of the facilities that these officials operate. Added to these rumours are suggestions and suspicions of corruption surrounding the plants. According to interviews with representatives of ENGOs, ordinary citizens, and people who were familiar with the starting-up process of incinerators, such facilities can only operate with official government permissions (Interviews, 2017). These permissions generally are granted based on written documentation regarding the processes within the plants as provided by the company that operates the facility. This documentation is inspected and dealt with by the municipal or state authority that is responsible for granting permission to operate. The process of receiving permission does not include on-the-spot inspections of the plants or unannounced visits by the environmental department(s) responsible for the area. In many cases, as it turns out, the incineration processes that have received official permission bear no semblance to the actual work that takes place within the plants.

The Tianjin Explosion of 2015

One incident that was regularly referenced by the persons I spoke with in the spring of 2017 had no connection whatsoever to garbage incineration: the 'Tianjin explosion'. This has become the catch phrase covering the steady flow of food safety scandals, industrial plant accidents, and the continuing pollution of water, air, and land that has taken place in recent times (He et al., 2013).

The explosion took place in a container storage station operated by Ruihai Logistics Co. Ltd. in the port of Tianjin on 12 August 2015, killing 173 people and wounding hundreds more. The damage wrought by the blast was enormous, destroying more than 300 buildings and almost 12,500 cars, totalling almost 6.9 billion yuan worth of damages (Fu, Wang, and Yan, 2016). The causes of the explosion were manifold, but can best be summarized as poor or absent inspection and audit practices by government agencies, which resulted in management deficiencies; a shocking lack of safety knowledge, awareness, and habits; illegal construction and operating issues; problems associated with risk control failures; and violations of national or industry standards (Chuang, 2015; Swiss RE, 2016). This was further exacerbated by a confusing overlap of responsibilities among the responsible administrative bodies at various levels, eventually resulting in the unlawful combined storage of chemical components in life-threatening quantities. Although the facility had, often fraudulently, obtained all the documents and permissions it needed to operate, it had also engaged in illegal activities that were not noticed by the authorities or against which no legal actions were undertaken (Chuang, 2015; Fu, Wang, and Yan, 2016). Such behaviour was helped further by the fact that the organization operating the facility doubled as its regulator, and that the persons involved in the permission-granting process had family ties to politicians at the highest levels of government (Chuang, 2015). In the period from 7 to 9 November 2016, 49 people, 25 government officials and 24 staff members of the companies involved were jailed for periods ranging from three years to life by ten separate courts; the Chairman of Ruihai Logistics was convicted of the illegal storage of hazardous materials, illegal business operations, causing incidents involving hazardous materials, and bribery. He received a death sentence with a two-year reprieve (*Xinhua*, 2016; Kennedy, 2016).

Lack of Faith

Reports about similar but smaller cases often circulate in the media and are widely shared through social media channels including Weibo and WeChat (Duggan, 2015). Some of these incidents generate considerable publicity in the official media as illustrations of how diligently the government is taking steps to eradicate them, yet they continue to occur. Many other stories still make the rounds, although they run the risk of being quickly deleted by the state censorship system. All revolve around collusion between government officials who turn political capital into economic gain, on one side, and operators of facilities with a deficient understanding of the law, on the other, in projects that earn them loads of money but at best give ordinary people nothing, and at worst endanger their lives (Balkan, 2012). Time and again, Beijingers cite these and similar examples of the illegal and corrupt management of plants involved in garbage recycling and incineration as a reason for their lack of faith. This also extends to a lack of trust in the environmental protection departments that are responsible for overseeing the plants. In the people's opinion, public welfare, the correct disposal of waste, improving the living environment, and the health of those working in the installations were not what concerned these departments or the industrial interest groups behind the incinerator initiatives that pushed their decisions through (Meng, 2010).

A concrete example mentioned by a number of people, including the representatives of ENGOs, was the Gao'antun Incinerator Facility in Chao-yang District, East Beijing. To try and quell rumours about the lack of safety procedures inside the incinerator and the potentially lethal side effects of its procedures for the workers, and by extension the people living in the neighbourhood, in 2000 the management ordered testing of the dioxin levels in workers' blood in the plant, with the intention of releasing the results to the public. Once the results became available and it turned out that dioxin levels were unacceptably high, they were switched with results taken from the management staff and presented as if they came from the workers (Interviews 2017; Interview with Huan You, 2017). Another story I heard frequently was that, despite all the assurances that the facilities operated in line with national and international norms, the actual emissions were considerably higher than international standards (Yang, 2013: 186; Fieldnotes, 2017).

Building trust

The main recurring complaint about the proposed construction of incinerators is that people felt left out of the loop. The decisions are made by the government, the companies involved in the construction, and scientific advisers (He et al., 2013). The public feels that they are not consulted during the Environmental Impact Assessment (EIA) process, to which they legally have access since the relevant legislation was enacted in 2003 (Li, Liu, and Li, 2012; Zhu et al., 2015); that they have no influence on the decision-making process; and suspect that crucial information is not disclosed to them, although the law entitles them to full disclosure (Xie, 2011; He et al., 2013; Wang, 2016; Interview with Green Beagle, 2017). Most of all, the people feel that they and their concerns and reservations are not taken seriously by the deciding bodies (Yang, 2013; Zhong, 2014; Zhong and Hwang, 2016). The government is aware of this popular antagonism and has come to understand that the people's confidence in both it and the project is essential (Liu et al., 2018). It has issued guidance to officials on methods by which to prepare the inhabitants of neighbourhoods for the arrival of a planned facility, which are much in line with the suggestions provided by my ENGO sources (for example, MOHURD, 2016).

Some incinerator plants have tried to develop a very open and above-board way of countering complaints, fears, and protests from the people living in their vicinity. These initiatives have had mixed results and have not been able to counter the public's general aversion. The Gao'antun Incinerator Facility in Chaoyang District is one of them. When Gao'antun started operations in 2008, the management erected a giant display screen on the grounds of the plant that recorded the real-time sulphur dioxide and nitrogen oxide emissions. Unfortunately, the screen was not clearly visible from outside of the plant, and these data were not accessible online. More worrying was the initial omission of data related to dioxin emissions, the secondary pollution released when burning plastics and other synthetic materials. This was later remedied, albeit with a time lag in the reporting (Watts, 2010; Johnson, 2013a). The Gao'antun management tries to further allay suspicions by opening the plant's doors, to make the processes taking place as visible as possible, by creating the impression that they take the fears and complaints of people living in the vicinity seriously, by organizing neighbourhood activities, etc. (Zheng S., 2017; Beijing Municipal Chaoyang Circular Economy Industry Park Management Centre, 2018b, 2018d). The Gao'antun plant is very active on Chinese social media, showcasing how the formerly waste and stench-ridden landfill has now turned into a veritable garden. The yearly postings on the company's WeChat account about viewing the blossoms in spring, a popular activity, are just one example of these public relation initiatives (Beijing Municipal Chaoyang Circular Economy Industry Park Management Centre, 2018c). Yet suspicions remain about this plant and the many others that are in operation across the country, particularly those in tier-two, -three, and -four cities (Wan, Chen, and Craig, 2015). Many see their public relations activities and postings as crude and simple attempts to whitewash nefarious activities and hoodwink more gullible residents, or to hide the true nature of what is happening inside the plants (Interviews, 2017).

Gao'antun garbage culture day trip

On her Weibo account, Lianpeng, a member of the ENGO Friends of Nature, critically reviewed a garbage culture event regularly organized by the Gao'antun plant; this review was republished on the Friends of Nature website (Lianpeng, 2017). In spring 2017, I also visited the Information Centre that features in her review and saw how a class of schoolchildren was educated (see also Beijing Municipal Chaoyang Circular Economy Industry Park Management Centre, 2018a). In general, Lianpeng believes that the event she participated in in 2014 was a failure. To begin with, it was not clear whether the visit was intended to show off the technology of the plant or to teach the visitors tips and techniques about reducing waste and classifying and/or separating it. The guide had a bored, annoying voice and presented the information in an unprofessional way. Her presentation was full of numbers and figures that showed how advanced the technology used at the plant is, underestimating the intelligence of the pupils. She also failed to mention the possibilities of harmless garbage disposal or provide useful information about how and why to recycle garbage. The visitors were surely impressed by the technology, but they did not feel connected to the whole process taking place. The district authorities (but not the plant management!) generously distributed many gifts to the visitors, but Lianpeng questioned the kindness and relevance of these gifts. Four of them were packed in plastic and most had little use in general. She thought it particularly weird that they were given fans, as the visit took place in autumn. She would have preferred more relevant and more appropriately packaged presents, such as little bags of fertilizer from the nearby landfill.

Lianpeng offered a number of suggestions to make the visit more worthwhile. The first was that the Information Centre should prepare lists of questions and tasks for the different types of groups that visit, and to make

Illustration 7.2   Display in the Circular Economy Information Centre at the Beijing Municipal Chaoyang Circular Economy Industry Park, showing the dioxin level of the incinerator emissions (right column), and comparing it with EU-levels and the standards for China/Beijing

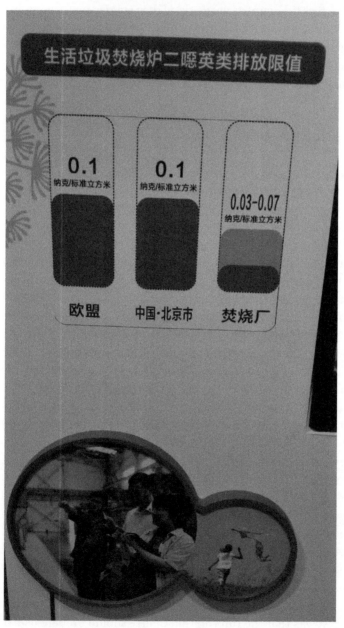

Author's photograph, 13 March 2017

sure that these materials are distributed before the visit takes place. Then, when the groups arrive, they should be shown around the landfill site first, and then the incineration plant. This will make it clear to the visitors why incineration has been chosen. Second, when visiting the incinerator, aside from introducing the technological aspects, more information should be provided that is related to the neighbourhood and society at large. This should include topics such as the safe distance between the incinerator plant and neighbouring residential communities, the safety system that is in place, and the monitoring system that controls the whole operation.

Third, the visit should include the waste composting field. Here, detailed information about the classification of kitchen waste can be provided. In Lianpeng's opinion, highlighting the difficulties that the sanitation workers experience when collecting kitchen waste, and the unpleasant smell in which they have to do their jobs, could persuade residents to separate the waste themselves.

The visits should be evaluated by designing questionnaires for the adults and quizzes for the children, with which they can win small prizes. The evaluation results can then serve as feedback for the guides.

These suggestions are certainly relevant, and some seem to have been adopted by the incinerator publicity department. Yet it should be noted that the Beijing Cultural Department and the Beijing Municipal Administration were the organization that contacted Lianpeng in response to her review, not the Beijing Municipal Chaoyang Circular Economy Industry Park Management Centre which appears to take care of external communications. During my visit to the Information Centre, a few years after Lianpeng's, I was struck by the motivated and lively way the guide did her job, even when most of the secondary-school pupils whose tour I joined were hardly interested. The displayed information also provided more than mere statistics and data, and some of the displays were interactive, attracting a lot of participation from the pupils (Illustration 7.2).

Popular opposition to incineration

Opposition to incineration and to the construction of other potentially polluting and environmentally hazardous factories such as PX (paraxylene) plants is one of the few things that can galvanize people into action. In many instances, however, such actions and protests do not evolve beyond the NIMBY level (Lee and Ho, 2014; Steinhardt and Wu, 2015; Johnson 2103a, 2013b; Wong, 2016; Zhu, 2017; Bondes and Johnson, 2017). People are only, and mainly, concerned about events taking place in their own neighbourhoods, and do not care about what happens in others. In short, as far as many are concerned, a principled opposition or resistance against incineration as such (Not-In-Anybody's-Back-Yard) has not yet emerged, although this situation is gradually changing (Bondes and Johnson, 2017; Interviews, 2017). Moreover, NIMBY protests are generally seen as emerging from the tension between the state and growing civil society interests and the demands from the expanding urban middle classes. Missing from these protests are other concerned parties, such as the people who are involved in the actual informal recycling (Tong and Tao 2016).

While ENGOs are hesitant about joining NIMBY protests in fear of encountering a political backlash that may threaten their existence, they do play a steadily growing role in providing these movements with the relevant legal knowledge needed to organize successful activities (Bondes and Johnson, 2017). They are also hard at work to increase the broader environmental awareness of the participants. This can be seen as an extension of their educational work. But as the Friends of Nature staff member pointed out, they are no match for a government that is determined to pursue the policies of incineration, whatever the costs.

8   Breaking the Waste Siege

'The City Besieged by Waste', Wang Jiuliang's 2011 documentary about the countless illegal garbage dumps surrounding Beijing, puts the problems that the disposal of MSW causes the city of Beijing, and by extension the rest of urban(ized) China, into a stark spotlight. The country faces the enormous task of preventing the urban areas and countryside alike from being buried in waste in the years to come. Solving the MSW problem would benefit the Chinese environment, the health of its people, and ultimately the rest of the world as well. In the preceding chapters, I have looked at a number of aspects of waste management that play a role in this important but extremely complicated question. I have also presented the views of a number of the participants in this process, all of whom need to be brought on board to find a possible solution.

Will China's cities drown in the waste produced by their inhabitants? That will probably not happen. Various national policies have been adopted, a number of structural decisions have been reached, and new trends of consciousness among the people have emerged. The effects of these developments will only be visible over a longer period of time, and they cannot be seen in isolation from events taking place in the rest of the world. The results may not turn out as we expect right now, but we will have to accept them for what they are.

Laws and regulations

China is known for the many stringent environmental laws it has adopted and promulgated over the past three decades. A steady stream of decisions, circulars, opinions, and other policy documents emanates almost daily from the central, provincial, municipal, and residential community level, each dealing with problems concerning the environment that urgently need to be solved. The government has expressed its commitment to creating a circular economy that will play a large role in the harmonious development it has also subscribed to. This harmonious development must balance the demand for resources with their continued availability and protect the environment at the same time. Until now, the only tangible and direct effect that the circular economy plan seems to have had on the lives of the people is the frenzy with which commercial parties have jumped onto the idea of the sharing economy. The dockless, shareable OfO, MoBikes, and others that entered Beijing in 2016 are the most visible examples. People initially embraced the idea of sharing a bicycle because it had health benefits, contributed to cutting down air pollution a little, and often was more convenient than using the overcrowded public transport options. But the support for shared bikes waned when navigating the mountains of left-behind bicycles on the sidewalks became problematic, when physically impaired people stumbled over them, and when the bikes turned out to be just another source of environmental pollution and solid waste, instead of the beginning of a solution.

Over the years, the country has entered and expressed explicit commitment to numerous global agreements to combat climate change. Yet all these efforts have produced very few concrete results until present. The Chinese environment is seen as one of the most polluted in the world. Official promises to take decisive steps and improve conditions have had little or no effect over time. Although systems of reward and punishment are built into the system, these have little influence.

The gap that often shows between government intentions and their implementation is well-known and has been researched extensively. It has given the Chinese government the reputation of only engaging in symbolic legislation and action. As I discuss in Chapter 2, the nature of the country and the structure of the political system as it functions at present are very much responsible for this situation. Because the central government is in charge of a very large and diverse nation and population, it is forced to formulate policies and laws in the broadest terms possible. By choosing to use vague and non-specific language in its official utterances, instead of providing narrowly defined targets and timeframes, the state intends to make them relevant for all. However, this allows local governments to argue that some or all of the promulgated laws and regulations in certain fields do not apply to them and to disregard the national benefit in favour of the local, and it allows individual government officials to pursue developments with hard targets like GDP that look well on their own performance reviews while avoiding softer targets that only may have positive effects in later years.

Relations between the political centre and the localities are formally clear and hierarchic, but in reality they are highly decentralized. Much political activity is marked by using connections, pulling rank, and other types of trade-offs. Given the complexity of the task of governing China, it is very difficult to bring local governments to heel without throwing the delicate framework of the governing system into disarray. The actions of local governments and the people working in them are inspected and evaluated closely and on a regular basis. Yet it has been, and continues to be, possible to fool the higher-ups by presenting model projects that conform to the rules, while continuing to operate other projects that are utterly and openly contravening those same rules. The central government engages in spot checks and temporary disciplinary campaigns to bring these practices to a stop, but these attempts do not have much effect beyond temporarily improving the behaviour of some local authorities.

Trust issues

Over the years, various opinion polls have demonstrated that the Chinese people have great trust in their government, as discussed in Chapter 7. This has confounded many scholars, mainly from the West, who were and are convinced that the population is restive, suffering under an authoritarian regime, and about to rise in revolution. The recorded high levels of popular trust concern only the central government; the actions and efficiency of local levels of government, on the other hand, are seen with extreme distrust, dissatisfaction, and suspicion. This creates the paradoxical situation that citizens complain openly about the (local) government not engaging in any action, while at the same time expressing their confidence that the (central) government will take the necessary steps to change the things that need to be changed. However, while the central government may have the authority to do so, it does not have the means to force local governments to comply with its decisions in the way they were intended. Central initiatives that would lead to the solution of a problem need to be put into operation by local administrative layers, and policies promulgated by the local governments, almost by definition, tend not to be accepted by the population because they are distrusted.

The problems and events surrounding the incineration of garbage is one of the ultimate expressions of this distrust. Over the course of the past two and the current Five Year Plans, the government has committed to solving the garbage problem by burning waste. Ideally, this burnt waste can be transformed into energy. Over the years, incinerator technology and techniques have been developed to extremely high levels of proficiency. The negative effects of incineration, including secondary dioxin pollution, can increasingly be kept to a minimum. From the global perspective, the Chinese embrace of waste incineration may be of great value, in terms of allowing the country to honour the international commitments it underwrote in aiming to stop or at least slow down climate change. China is one of the largest clients in the world when it comes to the acquisition of incineration technology, yet the

Chinese people remain unconvinced of the positive aspects of incineration. They point to recurrent large-scale industrial accidents, such as the Tianjin Explosion of 2015, to explain their rejection. While this explosion did not involve an incinerator facility, the factors that caused it to blow up also play an important role in waste burning. Industrial complexes like the Tianjin plant can only operate after applying for and receiving official permits from central and municipal government agencies that oversee whether safety practices and precautions are in place, environmental regulations are being followed, etc. The Tianjin case, as well as others, demonstrated that officials and entrepreneurs can conspire to engage in illegal activities that may bring them financial and political profits, but that have detrimental or even lethal effects on the people living in the vicinity of these projects. The cynicism of ordinary Chinese people regarding official corruption and the absence of care for the general wellbeing has reached such a level that they see a potential Tianjin explosion in every large-scale industrial plant. The steps that need to be taken in the planning and siting phases of an incinerator facility only occasionally follow the rules, as Johnson (2013a, 2013b) and others have demonstrated. People have the legal right to be informed, they have to be consulted in the context of the environmental impact assessment that needs to take place, etc., but in practice, their views and opinions are neglected. As a result, incineration meets with resistance, no matter whether this is in the form of a NIMBY protest against the fact that the people's investments in real estate plunge in value, or people actually taking a principled stand against the process.

The 13th Five Year Plan (2015-2020) calls for the construction of hundreds of incinerator facilities all over the country. While these plants will contribute to the ultimate disposal of MSW and will generate part of the energy that the country needs, the question is whether they will help in combatting the air pollution, specifically the PM2.5 pollution, that is currently at the centre of popular and government attention. The blue skies push in large urban areas has successfully increased the number of days with relatively low air pollution. It has demonstrated that when the government really puts its force behind an effort, results will follow. But the skies could only turn bluer than they were before because various industries were ordered to temporarily stop operating. No lasting policies have been enacted.

Air pollution from incineration is caused partly by the improper sorting of MSW before combustion. There are still too many plastic, paper, cardboard, glass, and other recyclable goods mixed in the waste stream. Simply burning away this MSW leads to secondary pollution, something many Beijingers are well aware of, as described in Chapter 3, and ENGOs are campaigning actively against it, as introduced in Chapter 6. Moreover, Chinese MSW is extraordinarily wet when compared with MSW elsewhere. This is the result of the large percentage of kitchen waste that is mixed in with the garbage before disposal by the residents. For this MSW to burn away as completely and cleanly as possible, it needs extra pre-treatment steps, for example by dehydrating the wet waste before incineration. Another option is to add combustion improvers like coal or kerosene to the waste. Such steps make the exhaust fumes more polluted and toxic and result in more wear-and-tear on the equipment. The net gains in energy that incineration can produce thus disappear. All of these solutions make the process more expensive than originally proposed, and more intrusive on the environment than the planners suggest.

Trust, as the government itself is well aware, needs to be improved. Popular distrust hampers development. It makes the choice for incineration highly contested and difficult to implement.

Front-end solutions

While incineration is a back-end solution to the MSW problem, many ENGOs urge that more attention be paid to changing behaviour at the front-end. As Chapter 6 showed, under the principles of the circular economy that call for consuming less and reusing and recycling more, ENGOs are committed to continuing to raise the awareness of the consumer-residents regarding these concepts and spur them into action. When it comes to recycling, the governments at the central and lower levels have put forward various rulings urging the citizenry to start engaging in classifying and separating the garbage it produces in ever-larger amounts. Often collaborating with ENGOs, street and residential committees have started educational pilot projects in residential communities since as early as 2000. The people on the bottom rungs of society and the ones at the very top tend to be more aware and more willing to change their recycling habits. The middle groups, however – the ones whose numbers have been expanding spectacularly over the past decades – are much less interested in doing so. Yet my sources told me, as presented in detail in Chapter 3, that they had no idea how to go about these tasks. Their apartments were too small and had no room to spare to store separated garbage. Nobody had told them which sort of garbage needed to be disposed of in which container. While the officials decided and planned and plotted courses of action, somehow they had neglected to inform the people who were supposed to adopt the desired behaviour about what they were supposed to do. Even when they were informed, it had no lasting effect over time. This is illustrated by the citizen remarks recorded during the doorstepping campaign undertaken by Dai et al. (2015) in Shanghai. The residents stated that they needed to be reminded every couple of weeks about the necessity and urgency of classifying and separating garbage. They had not sufficiently internalized the reasons behind such behaviour; they did not yet have the bigger picture in mind.

People are not saints. The preceding paragraph may create the impression that the citizens are completely left in the dark and are supposed to do things they have no knowledge about. This is not true; many are aware that environmental awareness and garbage separation are closely connected. The government has undertaken extensive education and propaganda campaigns to try and persuade the people to adopt new behaviours when it comes to disposing of their garbage. These campaigns have often taken recourse to the tested and trusted methods of continuous education and education-by-example, similar to the many political and other campaigns that have taken place in China over the past seven decades or so. The basic dialectical assumption underpinning these campaigns is that people want to change, as a natural result of being confronted with a behavioural example that is appealing to them. They compare their actions with the actions of the model they want to emulate. However, practice has shown that this educational effect does not take place as automatically and directly as the authorities have assumed over time. Indeed, not complying with behavioural stimuli can actually be seen as an act of resistance, as a way of expressing one's dissatisfaction. There are other reasons that explain the non-compliance of the people, or their unwillingness to adopt other behaviours. Many simply do not understand why they should change their old ways. They see the posters put up by the government(s), participate in the drives organized by the government(s) and the ENGOs, and all of this may contribute to their motivation to classify and separate their garbage. It does not, however, change their habits: the motivation to act is not translated into real action.

To bring this about, hands-on education is needed. As the ENGOs indicated in Chapter 6, the best results can be gained by training people rather than ordering them to do things through educational programmes and drives. Friends of Nature discovered that the residents of the communities they worked in tended to participate in the activities the ENGO set up. The 'greensleeves' introduced in Chapter 5 were also positive about the effects their educational work had on the actual behaviour of the residents, especially when they let their provision of information go hand-in-hand with the concrete example they set. Even so, both FoN and the 'greensleeves' reported that in some communities, the residents simply thought that these outside workers were there to do the classification and separation work for them.

The issue of trust is also pertinent to an individual's willingness to engage voluntarily in waste sorting. S/he must be absolutely convinced that the sorting work s/he has done actually makes a difference, and that the sorted garbage is treated in a sorted way. The municipal sanitation departments that are responsible for emptying the garbage containers and carting off the sorted garbage must prove beyond any doubt that they are acting as expected. They must avoid the possibility that containers with sorted and separated garbage end up in the trucks with the unsorted waste at all costs. This calls for very close attention to sanitation work and an attitude of openness that at the moment is still lacking.

It remains questionable whether appealing to individual willingness will lead to any results. Many residents I interviewed made it clear that waste and garbage are not their problem, but the government's. When the government tells them to take action, they will obey. They will grumble, for sure, and complain about others. But they made it very clear that the government has to act by promulgating rules and laws.

Rewards and penalties

Instead of merely calling upon people's better selves to encourage them to change their behaviour, other methods have proven to produce more results. In Chapter 3 and elsewhere, I referred to the need to adopt of a system of rewards and penalties to spur residents to start adopting and internalizing behaviour that will ultimately lead to less consumption and more systematic classification and separation of garbage. Previous research has shown that some sort of incentive is needed to change behaviour (Tang, Chen, and Luo, 2011: 869). At various moments in the recent past, municipal governments have experimented with, for example, pay-as-you-throw systems, i.e., making residents pay for the amount of garbage they threw away. These did have some positive effects on the total amount of waste that was discarded, but it also prompted some residents to devise ways to get rid of their garbage without having to pay for it. A clever solution adopted by some to evade waste payments was identifying neighbourhoods where the system was not in force and dumping their junk in the containers there. Others, however, simply dumped their waste wherever they thought they could get away with it. Note that such behaviour is not a uniquely Chinese phenomenon;

during experiments with similar systems in cities in the Western world comparable actions were recorded. This shows that a pay-as-you-throw system can only be effective when it is introduced comprehensively and covers the whole urban configuration. Moreover, it calls for a system of penalties for offenders and the creation of a supervising agency to ensure compliance. Public rewards for desired behaviour can further help the adoption of such a system.

Dai et al.'s doorstepping campaign is another strategy that produced initially promising results. It was obligation-free and therefore more difficult to enforce. Some of the participating residents, however, already let it be known that their increased awareness resulting from the campaign would need constant tending; they needed regular reminders of the behaviour that was required. Likewise, some made it clear that it was up to community management companies to organize the process of constantly raising residents' awareness; this was part of what they expected to receive in return for the service fees they paid. To a large extent, the positive results of Dai et al.'s experiment were influenced by the participation of a group of highly motivated young students who used face-to-face interactions to urge, but not force, residents to change their behaviours. Their idealistic commitment and high level of human quality greatly impressed the residents but did not fundamentally change the latter's behaviour.

A fairly recent approach meant to make citizens change their behaviours is the planned adoption of the Social Credit System (SCS), introduced briefly in Chapter 5. This System sets out to provide more transparency in the political, judicial, and economic domains, focusing on improving trustworthiness, something that is necessary for further development. The System is also part of a concerted effort to raise the general quality of the nation and the people. On the level of the citizens, this is supposed to happen by rating the behaviour of individuals. The narratives about quality, the people lacking it, and the ways to raise it, play a major role in present-day Chinese society. Good deeds are actions that are seen as beneficial to society; bad deeds harm the people. Earning credits for good deeds make life in general easier for the receiver; being penalized for bad behaviour can make it more difficult for the offender, ranging from being blocked from reserving tickets for high-speed trains to not being allowed to apply for a bank loan. In Hangzhou, where one of the pilot projects testing and evaluating the SCS took place, separating garbage was included among the desired behaviour that allows people to earn credits and improve their ratings. It is too early to tell whether the SCS in its final form will adopt waste-related behaviour among the categories that allow people to build up positive credit, but it is a promising new way of thinking about a system of rewards and penalties. From a technological point of view, it will not be difficult to integrate recycling-related data, which has already been collected by the various O2O companies that have developed smartphone applications to link junk producers with collectors, with the wider data systems that will be at the heart of the SCS.

Traditional practices, new approaches

The main reason residents hesitate to adopt new garbage-disposal behaviour is that they see no real need to do so. After the Beijing Municipal Resources Recycling Company started closing down its waste collection points in the 1990s, as detailed in Chapter 1, residents were stuck with their recyclables and forced to find ways to discard them themselves. Over time, informal waste collectors from the ranks of the migrant population filled the vacuum left behind by the Company. They grew into the army of recyclers that is presently responsible for much of the sorting of urban waste. A large number of them have moved from walking the streets collecting recyclables in a seemingly random manner to become more or less permanent fixtures in residential communities. As the 'guy downstairs', they have bought the right to and taken on the responsibility of sorting the residents' garbage, earning their living by selling the valuable junk they collect in the process. Their presence has not only succeeded in making more people committed to recycling, as I described in Chapter 3, but also makes the municipal classification and separation pilots and schemes seem much less urgent and relevant to the residents. The 'guy downstairs' certainly does not make the problems of waste and garbage go away, but s/he has taken on the job, thereby relieving the residents of their responsibilities. Management committees in communities that lack a 'guy downstairs' have adopted other solutions to their garbage problem, such as by contracting out the sorting process to favoured groups of informal collectors.

The State Council introduced the 'Internet Plus' policy in 2015 in an effort to make modern technology serve and revive various sectors of the economy. In the margins of this plan, 'Internet Recycle' made an entrance, where waste sorting, collecting, separation, and recycling activities were linked with the possibilities offered by information technology. Various companies have benefitted from government subsidies to develop new services in these fields. The Hong Chao Company introduced in Chapter 3 has come up with a modernized version of the 'guy downstairs'. Their Red Nests resemble the places these 'guys' work in, with the main difference that the Nests offer more sanitary working conditions, provide continuous education to residents, and make sure that the collected and sorted garbage is registered in a big data system and disposed of in a completely controlled manner. Hong Chao is convinced that this should facilitate the introduction of a pay-as-you-go system, which it sees as the most reliable way forward. Red Nest employees also engage in door-to-door pick-ups of recyclable junk.

Door-to-door pick-ups have been at the core of a number of other companies that adhere to the philosophy of 'online-to-offline', or O2O. Huishouge (Recycle Brother), developed by the Wuhan-based GEM Company, is one of the pioneers of the emerging O2O recycling sector. The companies have added offline activities to their online presence by developing smartphone applications to create networks of consumers who offer recyclables that they can then pick up. In some (parts of) districts in the Beijing area, Bangdaojia (Help at Home), Taoqibao and ZaiShenghuo (A new living), and more recently, LüMao (Green Cat) have started this kind of operation. The Bangdaojia app developed by the Incom Company plays a major role in Incom's efforts to harvest as much recyclable PET materials as it can to make its recycling factory in Shunyi District run full-time. It must supplement the catch of recyclable plastics that the company already harvests through the network of RVMs that Incom developed with government subsidies since the early 2010s. These RVMs are placed strategically in subway stations and other spots where the public congregates, but they do not yield enough plastic. Incom experiences too much competition from the informal waste-collecting field. Incom created Bangdaojia to build up a dependable network of consumers to offer recyclables, while also aiming to integrate its former competitors, the waste collectors, into its network by formally employing them. It is unclear how far the company has succeeded with this strategy.

Both Huishouge and Incom do more than pick up recyclable junk; they take care of the complete process of recycling and marketing the recycled resources. The nature of the operations of many other O2O recycling companies is less clear. Have they just taken over part of the already existing (informal) waste collection network? Were they competing with the established recycling companies? These questions remain unanswered. Taoqibao did not seem to be very active at all and, according to the persistent rumours circulating in 2017, the company has ceased to function. This impression is confirmed by the inactivity and disappearance of the Taoqibao accounts from the Internet. ZaiShenghuo, on the other hand, has branched out in other directions. Its app still has a tab for garbage collecting, but the thrust of its activities has shifted. Like an employment agency hiring out temporary employees, it now offers labour power for odd domestic and cleaning jobs as well as (wet) nurses and health workers, often collaborating with other job placement services.

None of my Beijing resident sources were familiar with the new O2O services. Some had seen and used the Incom RVMs, but they had never encountered the smartphone apps of Incom or any of the others. They had never heard of them and did not know anybody who used them. The O2O companies' publicity departments have apparently not been very successful in creating a buzz with their activities and services beyond the few reports in domestic and foreign media. It raises questions about the effectiveness and eventual success of such enterprises.

The informal sector

The informal sector, made up of the millions of migrants from the countryside who have found employment in the waste collecting and recycling system, needs to be taken into account in whichever method is chosen for the ultimate disposal of MSW. The waste sorters are generally credited with taking large quantities of recyclables out of the waste stream before their final disposal. This has saved the country major investments in exploring for and exploiting resources and prevented the mining and extraction of natural reserves, thus preserving part of the natural environment. Although all this has been relatively beneficial, there are critical voices that maintain that informal waste picking only takes the (more) valuable recyclables out of the waste. It does not solve the larger problem of disposal and, some argue, it contributes to secondary pollution by creating illegal waste dumps with unhygienic conditions.

Municipal governments, including Beijing's – which, as the capital, is under more (inter)national media scrutiny – have started to force the migrant population out of their territories. This caps the earlier moves to drive them out of municipal centres into suburban areas. Operating from the outskirts of the cities, waste sorting had already become more time consuming and therefore less well-paid. Who will sort the waste when all informal workers have been forced to return to their places of origin? Urbanites generally refuse to take on such demeaning, dirty, and low-paid work. Yet a way for the pre-sorting of waste must be found in order to make the garbage to be incinerated as combustible as possible. Joining existing, formal operations could provide a solution: informal workers could be absorbed into sanitation departments or O2O companies and continue their work. This allows for regulation of the business and for improving working conditions and making them more hygienic. Yet despite these positive changes, when I interviewed garbage workers, I discovered that many of them were not willing to give up their informal status.

Formalizing informal waste work will entail more than changing the designation of the workers involved. It also touches upon the *hukou* registration system that still keeps migrants from the countryside from settling in urban areas, and questions whether the benefits that urbanites enjoy will also be extended to the rest of the population. As China continues to urbanize, these uncertainties will need to be solved at some point.

Education and ENGOs

ENGOs have an important role to play in the existing strategies seeking a way out of the garbage problem. On the one hand, they give a voice to people who have been harmed in one way or the other by the effects of pollution and assist them with legal and compensation procedures. They assist informal workers and their families in creating a decent life. They are active in protecting animal and plant species. They serve as organizations where young people can hone their qualities as community leaders. But they also refrain from active participation in open contestations between the people and the authorities and keep as low a profile as possible. This is caused by the uncertainty of their status: they are embedded in the same administrative structure that they want to critically follow, and this ties their hands. Various political and organizational restrictions make it difficult for them to act in ways comparable to Western ENGOs. At best, they can play a role in the consultative processes that surround government decisions.

The most constructive contribution of ENGOs lies in the field of education and information provision. They have more knowledge of the ways in which social processes evolve within residential communities; they know what worries and problems live among the people; they have closer and more intimate relations with the residents. ENGOs can take on the role of an interface between the government and people. This will have a marked and beneficial effect on educational programmes designed to create and improve environmental awareness among the citizens. The present educational and propaganda materials designed and published by government agencies and committees just do not have the right tone, do not use the right phrases, lack the right nuances: they are official utterances, designed from a top-down perspective, are stodgy, and fail to strike a chord with their target groups.

Avenues for future research

This text has approached problems related to garbage classification and separation on the basis of the viewpoints and experiences of Beijingers in 2017. The situation in the countryside has been excluded completely, but according to some of my sources the pollution in rural areas is much more serious than in the cities. Many rural localities lack basic garbage disposal facilities and recycling opportunities, forcing the people to bury or burn waste themselves.

There are many topics related to the Chinese MSW problem that still beg for scholarly attention. One aspect that certainly deserves further investigation is the recycling of kitchen waste. The food and kitchen waste components in Chinese MSW are responsible for its wetness, making clean combustion problematic. This means that ways to make the treatment of kitchen waste more worthwhile must be explored. One of the problems encountered in biological or fermentation treatment is that there is no market for the compost that it produces. A first step in making its use more attractive is improving the quality of the compost by more rigorous sorting before treatment starts. Most importantly, however, the demand for compost must be stimulated. At the same time, solutions must be found for the sensory problem related to kitchen waste: the smell. Most residents refused to keep kitchen waste on their premises prior to disposal because they found it unhygienic and noxious. They were also convinced that the smell of rotting garbage had a negative influence on the way their quality was perceived by their neighbours, thereby harming their reputations. Entrepreneurs, scientists, and inventors should join hands with the government to develop methods and receptacles to neutralize these obstacles. Designing a scheme for the collection of kitchen waste for composting would make the waste mix for incineration less wet and easier to burn cleanly, and would avoid introducing expensive pre-treatment steps.

Other potential public-private partnerships need further study. The Hong Chao Red Nest version of the 'guy downstairs' phenomenon introduced in Chapter 3 is an example of how entrepreneurs who collaborate with the government can come up with solutions that may bring an end to some of the problems associated with waste separation. The O2O initiatives discussed in Chapter 2, the companies that developed smartphone applications to take over parts of the waste sorting and recycling process, are also examples of public-private partnerships. Most of them received government subsidies to develop parts of their operations. If the municipal governments nationwide persevere in their efforts to make it impossible for informal waste collectors to continue their work and to drive them out of the city and back to their rural places of origin, the O2O-companies would be in a favourable position to take over the waste-sorting activities. It would make perfect sense for them to formally join hands with the government departments responsible for waste disposal, as this would make the waste sorting work (look) more respectable, regulate the sector, and bring it under formal oversight and control. Further investigations are needed to find out under which conditions such cooperating bodies would be able to work.

Finally, the inclusion in the Social Credit System of the behavioural elements that play a part in the waste disposal programme is a topic that needs to be closely studied.

Appendix – Questionnaires used for research in 2017

Scrap collectors

Name
Gender
Age
Where from originally?
How long in Beijing?
How long in scrap collecting?
What did you do before? Back home? And here in the city?
Type of work now
- Scrap collector
- Collection point worker
- Scrap merchant
- Driver
- Other

Street collecting
- How did you start collecting?
- What are the benefits?
- How much do you make on a good day? On an average day? On a bad day?
- What is your total income per month?
- How about the stability of your income?
- What are the components of your income? (salary, bonus, or other form of income)

Where do you collect? Walking the street? Arrangement with community to pick up? What time do you start on an average day? Are there days with more scrap than others?

Are the numbers of scrap collectors getting smaller, generally speaking? Why? What are the reasons for that? Who is making work difficult for you? Which level of government makes your work most difficult?

O2O (Bangdaojia, ZaiShenghuo, TaoQiBao, or any others?)
–   Have you heard of them? How did you hear of them?
–   Are they a good alternative for collecting on the street?
–   Would you work for them? Have they approached you? Or have your
    friends/colleagues asked you to join?
–   Do you work for them?
    –   How did you start with them? Why this particular one?
    –   Only with one of them or with others too? Why more than one? How
        many hours for each one? Do they allow you to work for others too?
    –   What are the benefits of working for an O2O?
        –   More income? Regular income?
        –   Company uniform
        –   Tricycle? Truck?
        –   Tools and equipment
        –   Smartphone
        –   Scanner
    –   Have you been educated and trained by the company?
    –   Are there some conflicts between you and the company?
    –   Are there some conflicts between you and the app users?
    –   Are there conflicts of interest? Fights with other collectors? Fights
        with collectors from other companies?
    –   Do you want to do this job for a long time, and why?

Do you have a special relation with a garbage producer?
–   None
–   Regular
–   Frequent
–   An individual? A community? A company?
–   How did that relation start?
–   How do they get in touch with you?
–   What do you collect from them? For example, cardboard from a shop?
    Paper from a company?

After you have collected all day, where do you go to with your scrap? Always
the same merchant? Or different merchants for different types of scrap?
How did you build that relationship?

Garbage producers

Name
Gender
Age
Occupation
Family size/composition
Waste separation/garbage sorting at source?
– Yes/no
– Who takes care of it? You or your wife/husband?
– When, why, and how did you start separating?
– Do you think there is enough education/information about why separa-
  tion is good?
– What convinced you to separate?
– Something typically done by women or not?
– Does your home have enough space to separate?
– Suggestions for other separators?

MSW schemes
– The 'guy downstairs'
– Collection point
– App user
– O2O company of choice
– Only one?
– Why more?
– Why none?
– What are the comparative advantages of an O2O in your opinion?
– Special relation with recyclables (the value)?
– Special relation with recycler (the 'guy downstairs')?

O2O company questions

Company name
Date
Spokesperson(s)

Rank/occupation
1.   Operations
−   Scope
−   Active since
−   Why RVMs?
−   Collecting and/or recycling?
     −   Only recyclables or all SMW?
     −   Only PET?
−   Storage of recyclables
     −   Where? Collection points? Sorting stations?
     −   Are they run by the company or others?
     −   Where are they located?
−   Recycling
     −   Where does it take place?
     −   Own equipment/imported? Collaboration with equipment providers?
     −   How many production units?
     −   Turning out what sort of resource/product?
     −   Selling this product to whom?
−   Other company activities?

2.   Size (number of departments)
−   Names/functions

3.   Size (number of personnel)
−   Functions
−   Management
     −   Number
     −   M/F %
     −   Types of activities
−   Staff
     −   Number
     −   M/F %
     −   Types of activities

- Workers
  - Number
  - M/F %
  - Types of activities
  - Number/types of activities

4. National outreach (on the basis of information provided on the website)
- Which activities where?
  - Other plans?

5. International outreach (on the basis of information provided on the website)
- Which activities where?
  - Other plans?

6. Hiring outside scrap collectors
- Since when?
- Why did the company decide to hire them?
- How?
  - Active recruitment
    - How?
  - Word-of-mouth?
  - Other methods?
- Requirements for the job?
  - Is *Hukou* relevant?
  - Is age relevant?
  - Is education level relevant?
  - Does the company provide (on-the-job) training?
  - Apprenticeship?
- Only hiring scrap collectors? Or other related jobs as well?
  - Sorters?
  - Collection point / sorting station managers?
  - Drivers?
  - Recycle process workers?
  - App/website developers?
  - Others?

7. Structure and quality of employment
- Contract?
- Freelancers?

- Working hours?
- Supervision (of scrap collectors)?
    - Conflicts with other collectors?

8. Reimbursement
- (Base) salary
    - Approximately how much?
- Piece rate?
- Competitive bonus system?
- Insurance?
- Promotion opportunities?
- Penalties/discipline?

9. Gear
- Company uniform
    - What is it?
    - Is it being copied by other scrap collectors?
- (Hand) Carts, tricycles
- Vans
- Rented? Deposit?

10. Scope of activities within Beijing
- Where (most) active?
    - *Shequ*
    - Schools
        - Primary
        - Secondary
        - Tertiary
    - Offices
    - Commercial establishments
    - Shopping malls/shopping centres
    - Restaurants
    - Hotels
    - Neighbourhood
    - District
        - Which district produces most recyclables?
    - City-wide activities
    - Where of the above are most recyclables collected?

11. Collaboration with environmental government departments on activities?
–   *Xiaoqu*-level
–   Neighbourhood-level
–   District-level
–   City-level
–   What sort of collaboration?

12. Collaboration with labour departments for hiring personnel?
–   *Xiaoqu*-level
–   Neighbourhood-level
–   District-level
–   City-level
–   What sort of collaboration?

13. Collaboration with propaganda departments to co-organize/publicize activities?
–   *Xiaoqu*-level
–   Neighbourhood-level
–   District-level
–   City-level
–   What sort of collaboration?

14. Collaboration with environmental NGOs?
–   Which?
–   Why?
–   Types of collaboration
–   What sort of collaboration is relevant for the company?
–   What sort of collaboration is relevant for the ENGO?

15. Internet +
–   Bangdaojia app
    –   How many users?
    –   How successful in terms of recyclable collection?
–   How does it work in practice?
–   Is it city-wide?
–   Who develops/is responsible for the app?
–   Who is responsible for the website?

- Who is responsible for the communication strategy?
  - In collaboration with outside agencies (environmental departments, others, etc.)?
- Who is responsible for the contents?
  - In-house publicity department
- Who designs the educational content?
- Who designs the welfare content?
- The RVM-M2M system generates big data that are useful for environmental agencies and consumer goods producers. How can these data be put to other use?

General questions, wrapping up:
- Have the company results in terms of recyclable waste collected improved after the company started hiring outside collectors?
- Has the scrap flow been secured to satisfaction?
- Why did the company decide to branch out into other activities?
  - Smartphone repairs
  - Maintenance work in individuals' homes
  - Others
- How does the company hire these workers?
  - In collaboration with the Labour Bureau?
  - Are these jobs predominantly done by men?

NGO Interviews

1. There is a long history of recycling in Beijing. After Liberation, it was organized by the State. After 1980, the State withdrew and individuals took over. In recent years, migrant workers have become active as scrap collectors. I am interested in the recent emergence of so-called Online-to-Offline companies active in collecting recyclable waste in the city. Most of them operate through smartphone apps. Examples in Beijing are TaoQiBao, Bangdaojia, and Zaishenghuo; Huishouge is active in Shenzhen, Wuhan, and Tianjin. Through these apps, one can contact someone to come to one's home and pick up recyclables like plastic, glass, and paper in return for money or goods. The apps specify the value of the waste materials and the amount of money one receives in return.
   - Have you heard of these companies or do you know them? Are you familiar with their services? What do you think of them? Does your organization cooperate with them? If yes, in what way? If no, why not?

  – Do you think that these companies will be able to collect more recyclable waste than other methods?

  – The companies give the impression that they themselves take care of the recycling and that they do this in a way that protects the environment and the health of the people working in recycling. So not like what happens in many of the 'garbage' villages surrounding Beijing. Do you think they do a good job recycling? And that they do it in a healthier and environmentally friendly way?

  – Some of the companies have started to hire people who used to collect scrap in the streets to work for them instead. Have you heard of this practice, of hiring (migrant) workers to collect recyclables? Do you think this increases the efficiency of recyclable collecting? Do you think this is a good employment opportunity for these workers? Does it secure a stable income and improve their lives?

2. Are Beijing people well informed/educated when it comes to recycling? What do you think could be improved? How do you think the process of recycling could be improved? How is your organization involved in this process?

  – In 2000, the government told the people that they should start to separate their garbage. Do people separate their garbage? How is it done in practice? How could it be improved?

  – The O2O companies mentioned above all provide information/ education about recycling and garbage separation through their websites and apps. Are you familiar with the contents of this information/education? Do you approve of the way in which it is done, or do you have suggestions for improvements? Do you think that the information makes the consumers more aware of the need to produce less waste and dispose of it in a more environmentally friendly way?

  – Do you agree with the observation that elderly people are more committed to recycling than the younger generations (*80-hou, 90-hou*)? How could the younger generations be educated more effectively? What approach does your organization use?

3. Governments all over the world think that incineration of garbage is the best solution for the problem of dealing with waste. What is the attitude of the Chinese government towards incineration? What is the attitude of the people towards incineration?

  – Is the practice of incineration successful in China? In Beijing?

- Are incinerators properly equipped with the technology to let them operate in an environmentally friendly way?
- When garbage is not properly separated and too much kitchen waste ends up in the incinerator, the waste will be too wet, and it will become very difficult to incinerate the waste without adding external energy. What is the situation in Beijing like? Is the waste dry enough? Do incinerators here produce the amount of energy they have promised they would? What happens with the energy produced by incineration?
- All over the world, there are popular protests against incinerators. In Beijing too? Are these protests successful? In which way? Or are they not-in-my-backyard (NIMBY) protests, meaning that the plant will eventually be built somewhere else?
- Aside from consuming less, producing less waste, and recycling more, which suggestions does your organization have to deal with the waste problem? How do you communicate these suggestions to the public, to policy makers, etc.?

Municipal Government Department Interviews

1. In many of the interviews we have conducted with Beijing people from all walks of life, the respondents told us that education, propaganda, and information about recycling and garbage separation are the responsibility of the governments at all levels, from national to district to neighbourhood.
   - What are your ideas and insights about such education, propaganda, and information? What should it be like, what should it include, what should it stress most?

2. What types of educational programmes, propaganda initiatives, and information provisions have the Beijing Municipal government departments developed over the years? What are the most significant changes in the past few years regarding these activities? Have they become more frequent? Have they changed over time?
   - Can you tell us some more about the contents of these initiatives? What do you educate about? What forms can this take? What sort of materials do you prepare?

- Do you organize joint projects with ENGOs? What can and do they contribute to your efforts? And is it possible to tell which ENGOs you cooperate with?

3. It has come to my attention that you have been involved in garbage recycling movements in public schools. What were these movements like? How were they organized? What was your part in them?

4. Is it true that young people, i.e., the *80-hou, 90-hou,* and *00-hou,* need special attention when it comes to creating an awareness about the environment, recycling, and garbage separation? Are they less knowledgeable about these aspects than, for example, the *50-hou* and *60-hou?* Do they care less? Do they need special attention because they are the next generation, representing the future?
   - Do you think that creating awareness among the younger generations will have a positive effect on the recycling behaviour of their parents and grandparents?

5. Recently, so-called Online-to-Offline companies have become active in the chain of collecting recyclable waste in Chinese cities. Most of them operate through smart phone apps. Examples in Beijing are TaoQiBao, LüMao, Bangdaojia, and Zaishenghuo; Huishouge is another one that is active in Shenzhen, Wuhan, and Tianjin. Through these apps, one can contact someone to come to one's home and pick up recyclables like plastic, glass, and paper in return for money or goods. The apps specify the value of the waste materials and the amount of money one receives in return.
   - Have you heard of these companies or do you know them? Are you familiar with their services? What do you think of their work? Would they be mentioned in your educational work?
   - These O2O companies all provide information/education about recycling and garbage separation through their websites and apps. Are you familiar with the contents of this information/education? Do you approve of the way in which it is done, or do you have suggestions for improvements? Do you think that the information makes the consumers more aware of the need to produce less waste and dispose of it in a more environmentally friendly way? Would you consider joining hands with these companies when it comes to education, or organizing joint activities with them?

Bibliography

Agence France Press/JiJi (2018). 'China's waste import ban upends global recycling industry', *The Japan Times* 24 January 2018 (https://www.japantimes.co.jp/life/2018/01/24/environment/chinas-waste-import-ban-upends-global-recycling-industry/), accessed 1 February 2018

Albores, P., K. Petridis, P.K. Dey (2016). 'Analysing efficiency of Waste to Energy Systems: Using Data Envelopment Analysis in Municipal Solid Waste Management', *Procedia Environmental Sciences* 35, pp. 265-278

Alpermann, Björn (2013). 'Class, Citizenship, Ethnicity: Categories of Social Distinction and Identification in Contemporary China', in *Interdependencies of Social Categorisations*, ed. by Caniela Célleri, Tobias Schwarz, and Bea Wittger (Madrid: Iberoamericana), pp. 237-261

Anagnost, Ann (2004). 'The Corporeal Politics of Quality (*Suzhi*)', *Public Culture* 16:2, pp. 189-208

Anderson, Benedict (1991). *Imagined communities: reflections on the origin and spread of nationalism* (2nd ed.) (London: Verso)

Bakken, Børge (1994). *The Exemplary Society*, ISO-Rapport 9 (Oslo: Department of Sociology, University of Oslo)

Balkan, Elizabeth (2012). 'Dirty truth about China's incinerators', *China Dialogue* 7 April (https://www.chinadialogue.net/article/show/single/en/5024), accessed 15 February 2018

Battaglia, Gabriele (2017). 'Beijing Migrant Worker Evictions: The Four-Character Word You Can't Say Anymore', *South China Morning Post*, 3 December (http://www.scmp.com/week-asia/society/article/2122496/beijing-migrant-worker-evictions-four-character-word-you-cant-say), accessed 5 December 2017

Baudrillard, Jean (1998 [1970]). *The Consumer Society – Myths & Structures* (London: Sage Publications)

Beijing Evening News (2017). 'How Should We Recycle?' (垃圾分类，我们该怎么做), *Beijing Evening News* (北京晚报), 23 April 2017

Beijing Incom Resources Recovery Recycling Co. (no date). 'Company profile' (https://incomrecycle.en.alibaba.com/company_profile.html), accessed 29 September 2014

Beijing Municipal Chaoyang Circular Economy Industry Park Management Centre (2018a). 'The Classroom in the Park' (课堂进园区), 23 March 2018 (https://mp.weixin.qq.com/s/ftQ2pMdGW7Hkp-QI8q1hMg), accessed 1 May 2018

Beijing Municipal Chaoyang Circular Economy Industry Park Management Centre (2018b). 'Following the spirit of the MEE, we develop open activities' (响应环保部精神开展开放活动), 3 April 2018 (https://mp.weixin.qq.com/s/Sgp5CcAXOxhiFsDkNgNuAA), accessed 1 May 2018

Beijing Municipal Chaoyang Circular Economy Industry Park Management Centre (2018c). 'Spring in the park, the secret of flowers' (园区之春，花之密语), 18 April 2018 (https://mp.weixin.qq.com/s/JceZ-AaZ5qlMnsiP6unOxA), accessed 1 May 2018

Beijing Municipal Chaoyang Circular Economy Industry Park Management Centre (2018d). 'The Park is open to the Outside' (园区对外开放再迎小高峰), 23 April 2018 (https://mp.weixin.qq.com/s/f6YHT2jPckVfD3JtRCfjgQ), accessed 1 May 2018

Beijing Municipal Jingmeng Culture Communication Company Ltd. (北京市鲸梦文化传播有限公司) (2017). 'Xi Jinping cares about these six things' 习近平关心的这六件事 新华网 (https://www.youtube.com/watch?v=L3-WE8hVVuw), 14 March 2017, accessed 2 August 2017

Beijing Municipal Statistics Bureau (2017). '<Beijing in Numbers>: Household Garbage Classification – Are you ready?' (<数说北京>：生活垃圾分类您准备好了吗), BTV 5 June 2017 (http://www.bjstats.gov.cn/tjsj/ssbj/201706/t20170608_376034.html), accessed 2 August 2017

Beijing Municipal Urban Management Committee (北京市城市管理委员会) and Beijing Municipal Urban Management Committee Information Center (北京市城市管理委员会信息中心) (2011, 2012, 2013, 2014, 2015, 2016, 2017). *Urban Management Science & Technology* (城市管理与科技), vol. 13 (2011), 2-6; vol. 14 (2012), 1-6; vol. 15 (2013), 1-6; vol. 16 (2014), 1-5; vol. 17, 1-6 (2015); vol. 18, 1-6 (2016); vol. 19, 1-5 (2017)

Béja, Jean Philippe, Michel Bonnin, Feng Xiaoshuang and Tang Can (1999a). 'How Social Strata Come to Be Formed, Part 1', *China Perspectives* 23 (1999), pp. 28-41

Béja, Jean Philippe, Michel Bonnin, Feng Xiaoshuang and Tang Can (1999b). 'How Social Strata Come to Be Formed, Part 2', *China Perspectives* 24 (1999), pp. 44-54

Bergère, Marie-Claire (2002). 'China in the Wake of the Communist Revolution: Social Transformations, 1949-1966', in *China's Communist Revolutions – Fifty Years of The People's Republic of China*, ed. by Werner Draguhn and David S.G. Goodman (London, etc.: RoutledgeCurzon), pp. 98-123

Bhandari, Bibek (2017). 'Visually Impaired Say Shared Bikes Obstruct Sidewalks', *Sixth Tone*, 14 July 2017, accessed 15 July 2017 (http://www.sixthtone.com/news/1000516/Visually%20Impaired%20Say%20Shared%20Bikes%20Obstruct%20Sidewalks)

Bislev, Ane (2015). 'The Chinese Dream: Imagining China', *Fudan Journal of the Humanities and Social Sciences* 8, pp. 585-595

Blomsma, Fenna, and Geraldine Brennan (2017). 'The Emergence of Circular Economy – A New Framing Around Prolonging Resource Productivity', *Journal of Industrial Ecology* 21:3, pp. 603-614

Blumenthal, Eileen P. (1979). *Models in Chinese Moral Education: Perspectives from Children's Books* (Ann Arbor: UMI)

Boland, Alana, and Jiangang Zhu (2012). 'Public participation in China's green communities: Mobilizing memories and structuring incentives', *Geoforum* 43, pp. 147-157

Bondes, Maria, and Thomas Johnson (2017). 'Beyond Localized Environmental Contention: Horizontal and Vertical Diffusion in a Chinese Anti-Incinerator Campaign', *Journal of Contemporary China* 26:106, pp. 504-520

Brazil, Matthew (2018). 'China's Counterintelligence "Trinity" and Foreign Business', *China Brief* 18:5 (https://jamestown.org/wp-content/uploads/2018/03/PDF.pdf?x87069), accessed 30 March 2018

Broudehoux, Anne-Marie (2007). 'Spectacular Beijing: The Conspicuous Construction of an Olympic Metropolis', *Journal of Urban Affairs*, 29:4, pp. 383-399

Burch, Betty B. (1979). 'Models as Agents of Change in China', in *Value Change in Chinese Society*, ed. by Richard W. Wilson, Amy Auerbacher Wilson, and Sidney L. Greenblatt (New York N.Y.: Praeger Publishers), pp. 122-137

Callahan, William A. (2015). 'History, Tradition and the China Dream: socialist modernization in the World of Great Harmony', *Journal of Contemporary China* 24:96, pp. 983-1001

Carlson, Cameron (2017). 'Mobike receives Social Enterprise Award, criticism ensues', *China Development Brief* 27 March 2017 (http://www.chinadevelopmentbrief.cn/news/mobike-receives-social-enterprise-award-criticism-ensues/), accessed 3 April 2018

Central Committee of the Communist Party of China, State Council (2015). 'Opinions of the CPC Central Committee and the State Council on Further Promoting the Development of Ecological Civilization' (http://environmental-partnership.org/wp-content/uploads/download-folder/Eco-Guidelines_rev_Eng.pdf), accessed 1 March 2017

Chai, Jing (2015). 'Under the Dome' (https://www.youtube.com/watch?v=T6X2uwlQGQM, published on 1 March 2015), accessed 27 September 2015

Chan, Kam Wing (2009). 'The Chinese Hukou System at 50', *Eurasian Geography and Economics* 50:2, pp. 197-221

Chen, Liwen (2012). 'A Beijing Recycler's Life' (https://www.youtube.com/watch?v=hsfO8Nr4w Cc&feature=youtu.be), accessed 6 March 2017

Chen, Na (2017). 'Three Shared-Battery Companies Announce Investments on Same Day', *Sixth Tone*, 10 May 2017, accessed 11 May 2017 (http://www.sixthtone.com/news/1000172/Three%20 Shared-Battery%20Companies%20Announce%20Investments%20on%20Same%20Day)

Chen, Ximeng (2014). 'Dongxiaokou rubbish hub to be recycled into new urban development', *Global Times* 2 July 2014 (http://www.globaltimes.cn/content/868550.shtml), accessed 4 March 2017

Chen, Xudong, Yong Geng, and Tsuyoshi Fujita (2009). 'An overview of municipal solid waste management in China', *Waste Management* 30:4, pp. 716-724

Cheng, Anqi (2012), 'Waste not, want not', *ChinaDailyAsia* 19 January 2012 (http://www.china-dailyasia.com/life/2012-01/21/content_110868.html), accessed 19 March 2017

Cheng, Hefa and Yuanan Hu (2010). 'Municipal solid waste (MSW) as a renewable source of energy: Current and future practices in China', *Bioresource Technology* 101, pp. 3816-3824

Chew, Matthew (2003). 'The Dual Consequences of Cultural Localization: How Exposed Short Stockings Subvert and Sustain Global Cultural Hierarchy', *positions: east asia cultures critique* 11:2, pp. 479-509

Cherrier, Hélène (2010). 'Custodian behavior: A material expression of anti-consumerism', *Consumption, Markets and Culture* 13:3, pp. 259-272

Chin, Jill (2011). *Waste in Asia* (Singapore: Responsible Research Pte Ltd)

China Civilization Office, CCP Central Propaganda Department (2013). *Practice culture, establish a new practice public service advertisements* (讲文明树新风公益广告) (http://www.wenming. cn/jwmsxf_294/zggygg/), accessed 10 March 2017

China Youth News (2016). 'Beijing's 150,000 scavengers live near junkyards, Gangs rush to grab sites' (北京15万抬荒者全家驻垃圾场 帮派林立抢争地盘), *China Youth News*, 3 February 2016 (https://xw.qq.com/news/20160203007987/NEW2016020300798700), accessed 29 December 2016

Chinese Posters Foundation (2016). 'Strive to collect scrap metal and other waste materials!' *Chinese Posters* (http://chineseposters.net/posters/e15-329.php), accessed 16 December 2016

Cho, Mun Young (2012). '"Dividing the poor": State governance of differential impoverishment in northeast China', *American Ethnologist*, 39:1, pp. 187-200

Chuang (2015). 'The Tianjin Explosion: A Tragedy of Profit, Corruption, and China's Complicated Transition', *Chuang* (http://chuangcn.org/2015/08/tianjin-explosion/), 21 August 2015, accessed 1 March 2017

Chung, S.-S. and C.-S. Poon (2001). 'A comparison of waste-reduction practices and new environmental paradigm of rural and urban Chinese citizens', *Journal of Environmental Management* 62, pp. 3-19

Clapp, Jennifer (2002).Distancing of Waste: Overconsumption in a Global Economy', in *Confronting Consumption*, ed. by Tom Princen, Michael Maniates and Ken Conca (Cambridge: MIT Press, 2002), pp. 155-176

Collective Responsibility (2017). 'Sustainability Insights: Shanghai's Informal Waste Management, Shanghai, 2017', *Collective Responsibility* 2017 (http://www.coresponsibility.com/wp-content/ uploads/2017/06/China-Informal-Waste-Report.pdf), accessed 15 April 2018

Compilation and Translation Bureau, Central Committee of the Communist Party of China (2016). *The 13th Five-Year Plan for Economic and Social Development of the People's Republic of China 2016-2020*, Beijing 2016 (http://en.ndrc.gov.cn/newsrelease/201612/P020161207645765233498. pdf), accessed 3 December 2017

Cooper, Tim (2010). 'Recycling Modernity: Waste and Environmental History', *History Compass* 8/9, pp. 1114-1125

Creemers, Rogier (2017). 'Cyber China: Upgrading Propaganda, Public Opinion Work and Social Management for the Twenty-First Century', *Journal of Contemporary China* 26:103, pp. 85-100

Cwiertka, Katarzyna J., and Ewa Machotka (2018). 'Introduction', in *Consuming Life in Post-Bubble Japan – A Transdisciplinary Perspective*, ed. by Katarzyna J. Cwiertka and Ewa Machotka (Amsterdam: Amsterdam University Press), pp. 15-30

Dai, Jingyun, and Anthony J. Spires (2017). 'Advocacy in an Authoritarian State: How Grassroots Environmental NGOs Influence Local Governments in China', *The China Journal* 79, pp. 62-83

Dai, Y.C., M.P.R. Gordon, J.Y. Ye, D.Y. Xu, Z.Y. Lin, N.K.L. Robinson, R. Woodard, and M.K. Harder (2015). 'Why doorstepping can increase household waste recycling', *Resources, Conservation and Recycling* 102, pp. 9-19

Dalrymple, Theodore (2016). 'Great Britain, the litter bin of Europe', *City Journal*, Autumn (https://www.city-journal.org/html/trash-studies-14802.html), accessed 3 January 2017

Dauvergne, Peter, and Genevieve LeBaron (2013). 'The Social Cost of Environmental Solutions', *New Political Economy* 18:3, pp. 410-430

Davidson, Lincoln E. (2015). '"Internet Plus" and the Salvation of China's Rural Economy', *The Diplomat*, 17 July 2015 (http://thediplomat.com/2015/07/internet-plus-and-the-salvation-of-chinas-rural-economy/), accessed 8 September 2016

Davis, Deborah S. (2005). 'Urban Consumer Culture', *The China Quarterly* (2005), pp. 692-709

Davis, Mike (2006). *Planet of Slums* (London: Verso)

Debord, Guy (1994 [1967]). *The Society of the Spectacle*, transl. by Donald Nicholson-Smith (New York: Zone Books)

Dikötter, Frank (2006). *Exotic Commodities – Modern Objects and Everyday Life in China* (New York: Columbia University Press)

Dikötter, Frank (2010). *Mao's Great Famine – The History of China's Most Devastating Catastrophe, 1958-1962* (London: Bloomsbury)

Ding, Junjie (2016). 'A 30 cm. trash can and dignity that cannot be ignored' (30厘米的垃圾桶与不容忽视的尊严), 1 March 2016 (https://mp.weixin.qq.com/s/06DiVqgFJQlSy6S2dKLiDw), accessed 7 April 2017

Dirlik, Arif (1989). *The Origins of Chinese Communism* (New York, etc.: Oxford University Press)

Donald, Stephanie Hemelryk (1999). 'Children as Political Messengers: Art, Childhood, and Continuity', in *Picturing Power in the People's Republic of China – Posters of the Cultural Revolution*, ed. by Harriet Evans and Stephanie Donald (Markham, etc.: Rowman & Littlefield Publishers, Inc.), pp. 79-100

Donald, Stephanie Hemelryk (2011). 'Beijing Time, Black Snow and Magnificent Chaoyang – Sociality, Markets and Temporal Shift in China's Capital', *Theory, Culture & Society* 28:7-8, pp. 321-339

Dong, Madeleine Yue (1999). 'Juggling Bits – Tianqiao as Republican Beijing's Recycling Center', *Modern China* 25:3, pp. 303-342

Dong, Madeleine Yue (2003). *Republican Beijing – The City and Its Histories* (Berkeley: University of California Press)

Dorn, Thomas, Sabine Flamme and Michael Nelles (2012). 'A review of energy recovery from waste in China', *Waste Management & Research* 30:4, pp. 432-441

Douglas, Mary (1966). *Purity and Danger – An Analysis of the Concepts of Pollution and Taboo* (London, etc.: Routledge)

Downs, Mary, and Martin Medina (2000). 'A Short History of Scavenging', *Comparative Civilizations Review* 42, pp. 23-45

Drackner, Mikael (2005). 'What is waste? To whom? – An anthropological perspective on garbage', *Waste Management & Research* 23, pp. 175-181

Duggan, Jennifer (2015). 'Green China: Why Beijing Fears a Nascent Environmental Protest Movement', *TakePart*, 9 October 2015 (http://www.takepart.com/feature/2015/10/09/china-environmental-protest), accessed 12 March 2017

Dutton, Michael (1998). *Streetlife China* (Cambridge: Cambridge University Press)

Earthscan (2010). *Solid Waste Management in the World's Cities – Water and Sanitation in the World's Cities 2010* (London, etc.: United Nations Human Settlements Programme)

Eaton, Sarah, and Genia Kostka (2014). 'Authoritarian Environmentalism Undermined? Local Leaders' Horizons and Environmental Policy Implementation in China', *The China Quarterly* (2014), pp. 359-380

Eberhardt, Christopher (2015). 'Discourse on climate change in China: A public sphere without the public', *China Information* 29:1, pp. 33-59

Energy Research Institute of Academy of Macroeconomic Research and National Development and Reform Commission (2017). *China National Renewable Energy Centre, China Renewable Energy Outlook 2017* (Beijing: College of Environmental Sciences and Engineering, Peking University) (https://www.dena.de/fileadmin/dena/Dokumente/Themen_und_Projekte/Energiesysteme/CREO-2017-EN-20171113-1.pdf), accessed 15 April 2018

Engebretsen, Elisabeth L. (2013). 'Precarity, Survival, Change – China's Ant Tribes', *Suomen Antropologi: Journal of the Finnish Anthropological Society* 38:2, pp. 62-71

Ensmenger, Devona, Josh Goldstein, and Richard Mack (2005). 'Talking Trash: An Examination of Recycling and Solid Waste Management Policies, Economies, and Practices in Beijing', *East-West Connections: Review of Asian Studies* 5:1, pp. 115-133

Evans, David, Hugh Campbell and Anne Murcott (2013). 'A brief pre-history of food waste and the social sciences', *The Sociological Review* 60, pp. 5-26

Evans, Harriet (2014). 'Neglect of a neighbourhood: oral accounts of life in "old Beijing" since the eve of the People's Republic', *Urban History* 41:4,pp. 686-704

Ewoh, Andrew I.E., and Melissa Rollins (2011). 'The Role of Environmental NGOs in Chinese Public Policy', *Journal of Global Initiatives: Policy, Pedagogy, Perspective* 6: 1, pp. 45-60

Ezeah, Chukwunonye, Jak A. Fazakerley, and Clive L. Roberts (2013). 'Emerging trends in informal sector recycling in developing and transition countries', *Waste Management* 33, pp. 2509-2519

Farquhar, Mary Ann (1999). *Children's Literature in China – From Lu Xun to Mao Zedong* (Armonk, etc.: M.E. Sharpe)

Featherstone, Mike (1998 [1991]).*Consumer Culture & Postmodernism* (London: Sage Publications)

Fei, Fan, Lili Qu, Zongguo Wen, Yanyan Xue, and Huanan Zhang (2016). 'How to integrate the informal recycling system into municipal solid waste management in developing countries: Based on a China's case in Suzhou urban area', *Resources, Conservation and Recycling* 110, pp. 74-86

Fei, Xiaotong (1992). *From the soil: The foundations of Chinese society* (Berkeley: University of California Press)

Fernandéz-Stembridge, Leila, and Richard P. Madsen (2002). 'Beggars in the Socialist Market Economy', in *Popular China – Unofficial Culture in a Globalizing Society*, ed. by Perry Link, Richard P. Madsen, and Paul G. Pickowicz (Lanham: Rowman & Littlefield), pp. 207-230

Fong, Vanessa (2004). *Only Hope: Coming of Age Under China's One-child Policy* (Stanford: Stanford University Press)

Fraser, David (2000). 'Inventing Oasis – Luxury Housing Advertisements and Reconfiguring Domestic Urban Space in Shanghai', in *The Consumer Revolution in Urban China*, ed. by Deborah S. Davis (Berkeley: University of California Press), pp. 25-53

Friends of Nature (Ziran zhi you) (2013). *Report on the 2012 Investigation of Garbage Separation Pilot Project in Beijing Residential Communities* (2012 年北京垃圾分类试点小区调研报告) (March 2013)

Friends of Nature (2017a). 'Without retreat, the "Thirteenth Five-Year Plan" needs to move from waste disposal to waste management' (不进反退，"十三五"需要从垃圾处理彻底迈向垃圾管理), 11 May 2017 (http://www.fon.org.cn/index.php?option=com_k2&view=item&id=7008:2017-05-11-15-11-06&Itemid=201), accessed 1 May 2018

Friends of Nature (2017b). 'Environmental Protection Organizations's Overall Suggestions on the "Thirteenth Five-Year Plan" for the Construction of National Treatment Facilities for Harmless Domestic Waste (Consultation Draft)' (环保社会组织关于《"十三五"全国城镇生活垃圾无害化处理设施建设计划（征求意见稿）》的总体建议), 12 May 2017 (http://www.fon.org.cn/index.php?option=com_k2&view=item&id=10566:2017-05-12-08-14-53&Itemid=111), accessed 1 May 2018

Fu, Gui, Jianhao Wang, and Mingwei Yan (2016). 'Anatomy of Tianjin Port Fire and Explosion: Process and Causes', *Process Safety Progress* 35:3, pp. 216-220

Fukuyama, Francis (1995). *Trust: The social virtues and the creation of property* (London: Penguin)

Fung Business Intelligence Centre, Secretariat of the Expert Committee of the China General Chamber of Commerce (ECCGCC), China Business Herald Research Institute (2016). *"Internet Plus" initiative drives e-commerce expansion; mobile commerce and rural e-commerce high on the agenda*, Beijing (https://www.fbicgroup.com/sites/default/files/Ten%20Highlights%20of%20China%E2%80%99s%20Commercial%20Sector%202016_03.pdf), accessed 8 September 2016

Gamble, Sidney D., and John Stewart Burgess (c. 1921). *Peking, a social survey conducted under the auspices of the Princeton University Center in China and the Peking Young Men's Christian Association* (New York: George H. Doran Company)

Gao, Qin, and Fuhua Zhai (2017). 'Public Assistance, Economic Prospect, and Happiness in Urban China', *Social Indicators Research* 132:1, pp. 451-473

Gao, Ruge (2013). 'Rise of Environmental NGOs in China: Official Ambivalence and Contested Messages', *Journal of Political Risk* 1:8 (http://www.jpolrisk.com/rise-of-environmental-ngos-in-china-official-ambivalence-and-contested-messages/), accessed 28 April 2018

Garrett, Shirley S. (1974). 'The Chambers of Commerce and the YMCA', in *The Chinese City Between Two Worlds*, ed. by Mark Elvin and G. William Skinner (Stanford: Stanford University Press), pp. 213-238

Geall, Sam (2015a). 'Interpreting ecological civilisation (part one)', *China Dialogue* (https://www.chinadialogue.net/article/show/single/en/8018-Ecological-civilisation-vision-for-a-greener-China-part-one-), 6 July 2015, accessed 17 March 2018

Geall, Sam (2015b). 'Interpreting ecological civilisation (part two)', *China Dialogue* (https://www.chinadialogue.net/article/show/single/en/8027-Interpreting-ecological-civilisation-part-two-), 8 July 2015, accessed 17 March 2018

Geall, Sam (2015c). 'Interpreting ecological civilisation (part three)', *China Dialogue* (https://www.chinadialogue.net/article/show/single/en/8038-Interpreting-ecological-civilisation-part-three-), 10 July 2015, accessed 17 March 2018

Gille, Zsuzsa (2010). 'Actor networks, modes of production, and waste regimes: reassembling the macro-social', *Environment and Planning A* 42, pp. 1049-1064

Gillin, Brandon (2011). 'Keeping up with Chinese Consumerism: Offsetting China's Individually Generated Garbage with Regulatory and Social Mechanisms', *Vermont Journal of Environmental Law* 13, pp. 69-96

Goh, Esther C.L. (2009). 'Grandparents as childcare providers: An in-depth analysis of the case of Xiamen, China', *Journal of Aging Studies* 23, pp. 60-68

Goldman, Jasper (2003). *From Hutong to Hi-Rise: Explaining the Transformation of Old Beijing, 1990-2002* (unpublished master's thesis, Massachusetts Institute of Technology, Department of Urban Studies and Planning, September 2003)

Goldsmith, Elizabeth B., and Ronald E. Goldsmith (2011). 'Social influence and sustainability in households', *International Journal of Consumer Studies* 35, pp. 117-121

Goldstein, Joshua (2006). 'The Remains of the Everyday: One Hundred Years of Recycling in Beijing', in *Everyday Modernity in China*, ed. by Madeleine Yue Dong and Joshua Goldstein (Seattle: University of Washington Press), pp. 260-302

Goldstein, Joshua, and others (2011). 'The City Besieged by Garbage: Politics of Waste Production and Distribution in Beijing', UC Berkeley, 11 April 2011 (https://www.youtube.com/watch?v=BGPJSPOjywM), accessed 22 March 2012

Goldstein, Joshua (2013). 'How the World's Trash (Including Yours) Ends Up in China's Rivers', USC 18 September 2013, University of Southern California Dornsife, East Asian Studies Center (https://www.youtube.com/watch?v=pgx5mSTvt9U&feature=youtu.be&list=PLEAC41A08E1EE341A), accessed 30 May 2016

Goldstein, Joshua (2016). 'China's Changing Waste-Scapes', CSUSB Modern China Lecture Series, 17 May 2016 (https://www.youtube.com/watch?v=FhqkiXT_nNM), accessed 5 January 2017

Goldstein, Joshua (2017). 'Just how "wicked" is Beijing's waste problem? A response to "The rise and fall of a 'waste city' in the construction of an 'urban circular economic system': The changing landscape of waste in Beijing" by Xin Tong and Dongyan Tao', *Resources, Conservation and Recycling* 117, pp. 177-182

Gordon, Micheil (2013). *A framework to allow intervention design to increase food waste recycling in an urban community in Shanghai* – MPhil Thesis – 2013

Gow, Michael (2017). 'The Core Socialist Values of the Chinese Dream: towards a Chinese integral state', *Critical Asian Studies* 49:1, pp. 92-116

Grano, Simona A., and Yuheng Zhang (2016). 'New channels for popular participation in China: The case of an environmental protection movement in Nanjing', *China Information* 30:2, pp. 165-187

Greenfield, Adam (2018). 'China's Dystopian Tech Could Be Contagious', *The Atlantic* 14 February 2018 (https://www.theatlantic.com/technology/archive/2018/02/chinas-dangerous-dream-of-urban-control/553097/), accessed 16 February 2018

Gregson, Nicky, Alan Metcalfe, Louise Crewe (2007). 'Identity, mobility and the throwaway society', *Environment and Planning D: Society and Space* 25, pp. 682-700

Griffiths, James (2014). 'Here's why so many elderly Chinese are collecting trash', *GlobalPost* 16 July 2014 (http://www.pri.org/stories/2014-07-16/here-s-why-so-many-elderly-chinese-are-collecting-trash), accessed 24 November 2016

Gu, Edward X. (2001). 'Dismantling the Chinese mini-welfare state? Marketization and the politics of institutional transformation, 1979-1999', *Communist and Post-Communist Studies* 34, pp. 91-111

Guang, Lei (2005). 'Guerrilla Workfare: Migrant Renovators, State Power, and Informal Work in Urban China', *Politics & Society* 33:3, pp. 481-506

Guo, Ying (2016). 'Internet adds spring to China's circular economy', *Xinhuawang* (http://news.xinhuanet.com/english/2016-01/19/c_135023112.htm), 19 January 2016, accessed 28 August 2016

Gupta, Alok (2017). 'China's Internet giants look to go green ahead of "Singles' Day"', *CGTN.com*, accessed 11 November 2017 (https://news.cgtn.com/news/30516a4d35597a6333566d54/share_p.html)

Harvey, David (2003). *The New Imperialism* (Oxford, etc.: Oxford University Press)

Harvey, David (2005). *A Brief History of Neoliberalism* (New York, etc.: Oxford University Press)

He, Guizhen, Arthur P.J. Mol, Yonglong Lu (2012). 'Trust and Credibility in Governing China's Risk Society', *Environmental Science and Technology* 46, pp. 7442–7443

He, Guizhen, Arthur P.J. Mol, Lei Zhang, Yonglong Lu (2013). 'Public participation and trust in nuclear power development in China', *Renewable and Sustainable Energy Review* 23, pp. 1-11

Heberer, Thomas, and Anja Senz (2011). 'Streamlining Local Behavior Through Communication, Incentives and Control: A Case Study of Local Environmental Policies in China', *Journal of Current Chinese Affairs* 40:3, pp. 77-112

Hedrick, Lizzie (2016). 'Waste not, want not – the transformation of China's underground recycling industry', *USC News* 16 March 2016 (https://news.usc.edu/92924/waste-not-want-not-the-transformation-of-chinas-underground-recycling-industry/), accessed 23 August 2016

Hellmann, Kai-Uwe, and Marius K. Luedicke (2018). 'The Throwaway Society: a Look in the Back Mirror', *Journal of Consumer Policy* 41, pp. 83-87

Henriot, Christian (2013). 'Street Culture, Visual Fragments and Everyday Life: Narrating Peddlers in Shanghai Modern', in *Visualising China, 1845-1965 – Moving and Still Images in Historical Narratives*, ed. by Christian Henriot and Wen-hsin Yeh (Leiden: Brill), pp. 93-128

Hildebrandt, Timothy (2011). 'The Political Economy of Social Organization Registration in China, *The China Quarterly* (2011), pp. 970-989

Ho, Peter (2007). 'Embedded Activism and Political Change in a Semiauthoritarian Context', *China Information* 21:2, pp. 187-209

Ho, Peter (2008). 'Introduction', in *China's Embedded Activism – Opportunities and constraints of a social movement*, ed. by Peter Ho and Richard Edmonds (Routledge), pp. 1-19

Hongchao Enviro-Tech [红巢环保技] (no date). *Comprehensive plan for the disposal of urban garbage* [城市生活垃圾处理综合解决放案], no place, Hongchao Enviro-Tech

Hoornweg, Dan, Philip Lam, and Manisha Chaudhry (2005). 'Waste Management in China: Issues and Recommendations', *Urban development working papers* no. 9 (Washington, DC: World Bank)

Hoornweg, Daniel, and Perinaz Bhada-Tata (2012). 'What a Waste – A Global Review of Solid Waste Management', *World Bank Urban Development Series Knowledge Papers No. 15* (Washington: World Bank Urban Development & Local Government Unit)

Hooton, G.L.V. (1955). 'The Planning Structure and the Five Year Plan in China', in *Contemporary China*, ed. by E. Stuart Kirby (London: Hong Kong University Press), pp. 92-105

Hou, Yunlong (2015). '"Made in China 2025" roadmap surfaced, prioritize development in 10 areas' ("中国制造2025"路线图浮出水面 重点发展10大领域), *People.com.cn* (http://politics.people.com.cn/n/2015/0326/c1001-26751532.html), 26 March 2015, accessed 10 January 2018

Hsu, Jennifer Y.J., and Reza Hasmath (2014). 'The Local Corporatist State and NGO Relations in China', *Journal of Contemporary China* 23:87, pp. 516-534

Hsu, Jennifer Y.J., Carolyn L. Hsu, and Reza Hasmath (2017). 'NGO Strategies in an Authoritarian Context, and Their Implications for Citizenship: The Case of the People's Republic of China', *Voluntas* 28, pp. 1157-1179

Hu, Anning (2015). 'A loosening tray of sand? Age, period, and cohort effects on generalized trust in Reform-Era China, 1990-2007', *Social Science Research* 51, pp. 233-246

Hu, Jieren, Yue Tu and Tong Wu (2018). 'Selective Intervention in Dispute Resolution: Local Government and Community Governance in China', *Journal of Contemporary China* 27:111, pp. 423-439

Huang, Chuanhui (2015). *Migrant Workers and the City: Generation Now* (中国新生代农民工), transl. by Anna Beare (Beijing: Zhongyi chubanshe)

Huang, Haifeng (2016). 'Personal Character or Social Expectation: a formal analysis of "suzhi" in China', *Journal of Contemporary China* 25:102, pp. 908-922

Huang, Xuelei (2016). 'Deodorizing China: Odour, ordure, and colonial (dis)order in Shanghai, 1840s-1940s', *Modern Asian Studies* 50:3, pp. 1092-1122

Huishouge (2016). 'Congratulations to Huishouge's Successful Selection as one of the "One Hundred Best Practices of China's Internet Plus Movement"' (祝贺回收哥成功入选<中国 '互联网+' 行动百佳实践>), 14 December 2016 (https://mp.weixin.qq.com/s/xcyzz3eKzZT6piNNQKR_Ow), accessed 29 December 2016

Huishouge (2017a). 'Environmental Drive Classroom Unlocks the Mystery of Japanese Garbage Classification' (环保运动讲习所之解开日本垃圾分类之谜), 1 January 2017 (https://mp.weixin.qq.com/s/HsnJdNbL8tCL6eBPxC4Znw), accessed 11 January 2017

Huishouge (2017b). 'Do not forget your original intention, carry on forward' (不忘初心，继续前进), 4 January 2017 (https://mp.weixin.qq.com/s/7KEisHz6Zv7Y7Sb1rHrX9A), accessed 5 January 2017

Hunwick, Robert Foyle (2015). 'Scrap Empire', *The World of Chinese* 24 January 2015 (http://www.theworldofchinese.com/2015/01/scrap-empire/), accessed 24 March 2017

Hurst, William, and Kevin J. O'Brien (2002). 'China's Contentious Pensioners', *The China Quarterly* (2002), pp. 345-360

Inverardi-Ferri, Carlo (2017). 'Commons and the Right to the City in Contemporary China', *Made in China* 2:2, pp. 38-41

Jacka, Tamara (2009). 'Of Quality, Cultivating Citizens: Suzhi (Quality) Discourse in the PRC', *positions: east asia cultures critique* 17:3, pp. 523-535

Jian, Guo, Yongyi Song, and Yuan Zhou (2006). *Historical Dictionary of the Chinese Cultural Revolution* (Lanham, etc.: The Scarecrow Press, Inc.)

Jiang, Yuan, Kang Muyi, Liu Zheng, and Zhou Yanfang (2003). 'Urban garbage disposal and management in China', *Journal of Environmental Sciences* 15:4, pp. 531-540

Johnson, Thomas (2013a). 'The Politics of Waste Incineration in Beijing: The Limits of a Top-Down Approach?', *Journal of Environmental Policy & Planning* 15:1, pp. 109-128

Johnson, Thomas (2013b). 'The Health Factor in Anti-Waste Incinerator Campaigns in Beijing and Guangzhou', *The China Quarterly* (2013), pp. 356-375

Kanthor, Rebecca (2015). 'Chinese firm takes recycling into the "smart" era', *Plastic News China*, 8 July 2015 (http://www.plasticsnews.com/article/20150708/NEWS/150709927/chinese-firm-takes-recycling-into-the-smart-era), accessed 23 August 2016

Kao, Shih-yang (2011). 'Beijing Besieged by Garbage – Photographing "Year One": Wang Jiuliang and the Reign of Garbage', *Cross-Currents: East Asian History and Culture Review* 1 (https://cross-currents.berkeley.edu/e-journal/photo-essay/beijing-besieged-garbage), accessed 12 November 2015

Kao, Shih-yang, and George C. S. Lin (2018). 'The Political Economy of Debris Dumping in Post-Mao Beijing', *Modern China* 44:3, pp. 285-312

Kaplan, Richard L (2012). 'Between mass society and revolutionary praxis: The contradictions of Guy Debord's *Society of the Spectacle*', *European Journal of Cultural Studies* 15:4, pp. 457-478

Kennedy, John James, and Dan Chen (2018). 'State Capacity and Cadre Mobilization in China: The Elasticity of Policy Implementation', *Journal of Contemporary China* 27:111, pp. 393-405

Kennedy, Merrit (2016). 'China Jails 49 Over Deadly Tianjin Warehouse Explosions', *National Public Radio* 9 November 2016 (https://www.npr.org/sections/thetwo-way/2016/11/09/501441138/china-jails-49-over-deadly-tianjin-warehouse-explosions), accessed 1 March 2017

Kickbusch, Ilona (2007). 'Health Governance: The Health Society'. In *Health and Modernity – The Role of Theory in Health Promotion*, ed. by David V. McQueen and Ilona Kickbusch (New York: Springer Science+Business Media), pp. 144-161

Kipnis, Andrew (2006). 'Suzhi: A Keyword Approach', *The China Quarterly* (2006), pp. 295-313

Kipnis, Andrew (2011). 'Subjectification and education for quality in China', *Economy and Society* 40:2, pp. 289-306

Kostka, Genia (2014). 'Barriers to the Implementation of Environmental Policies at the Local Level in China', *Policy Research Working Paper 7016* (Washington DC: World Bank Groups)

Kostka, Genia, and Arthur P.J. Mol (2013). 'Implementation and Participation in China's Local Environmental Politics: Challenges and Innovations', *Journal of Environmental Policy & Planning* 15:1, pp. 3-16

Kostka, Genia, and Chunman Zhang (2018). 'Tightening the grip: environmental governance under Xi Jinping', *Environmental Politics* 27:5, pp. 769-781

Kuhn, Berthold (2016). 'Sustainable Development Discourses in China', *Journal of Sustainable Development* 9:6, pp. 158-167

Kuhn, Robert Lawrence (2013). 'Xi Jinping's Chinese Dream', *International Herald Tribune* 4 June 2013 http://www.nytimes.com/2013/06/05/opinion/global/xi-jinpings-chinese-dream.html, accessed 29 January 2014

Lam, Bourree (2016). 'The Anthropologist in the Landfill', *The Atlantic* 31 March 2016 (http://www.theatlantic.com/business/archive/2016/03/landfill-anthropologist/476121/)

Lan, Jing, Yuge Ma, Dajian Zhu, Diana Mangalagiu, and Thomas F. Thornton (2017). 'Enabling Value Co-Creation in the Sharing Economy: The Case of Mobike', *Sustainability* 9:1504

Landsberger, Stefan (1995). *Chinese Propaganda Posters – From Revolution to Modernization* (Amsterdam: The Pepin Press)

Landsberger, Stefan (1993). 'Role Modelling in Mainland China During the "Four Modernizations" Era: The Visual Dimension', in *Norms and the State in China*, ed. by Chun-chieh Huang and Erik Zürcher (Leiden: E.J. Brill), pp. 359-376

Landsberger, Stefan (2009). 'Harmony, Olympic Manners and Morals – Chinese Television and the "New Propaganda" of Public Service Advertising', *European Journal of East Asian Studies* 8:2, pp. 331-355

Landsberger, Stefan (2014). 'Dreaming the Chinese Dream: How the People's Republic of China moved from revolutionary goals to global ambitions', *History, Culture, Modernity* 2:3, pp. 245-274

Landsberger, Stefan (2018). 'China Dreaming – Representing the Perfect Present, Anticipating the Rosy Future', in *Urbanized Interfaces: Visual Arts, Representations and Interventions in Contemporary China*, ed. by Minna Valjakka and Wang Meiqin (Amsterdam: Amsterdam University Press), pp. 147-177

Lanza, Fabio (2018). 'A City of Workers, a City for Workers? Remaking Beijing Urban Space in the Early PRC', in *China: A Historical Geography of the Urban*, ed. by Yannan Ding, Maurizio Marinelli, and Xiaohong Zhang (Palgrave Macmillan), pp. 41-65

Larson, Christina (2010). 'Liang Congjie: The Godfather of China's Green Movement', *The Atlantic* 30 October 2010 (https://www.theatlantic.com/international/archive/2010/10/liang-congjie-the-godfather-of-chinas-green-movement/65442/), accessed 28 April 2018

Laskai, Lorand (2018). 'Why Does Everyone Hate Made in China 2025?', *Council on Foreign Relations* 28 March 2018 (https://www.cfr.org/blog/why-does-everyone-hate-made-china-2025), accessed 30 March 2018

Lee, Kingsyhon, and Ming-sho Ho (2014). 'The Maoming Anti-PX Protest of 2014', *China Perspectives* 3 (2014), pp. 33-39

Lefebvre, Henri (1995 [1962]). *Introduction to Modernity – Twelve Preludes September 1959-May 1961*, translated by John Moore (London: Verso)

Lefebvre, Henri (2002 [1961]). *Critique of Everyday Life VOLUME II – Foundations for a Sociology of the Everyday*, translated by John Moore (London: Verso)

Li, Bingqin (2006). 'Floating Population or Urban Citizens? Status, Social Provision and Circumstances of Rural-Urban Migrants in China', *Social Policy and Administration* 40:2, pp. 174-195

Li, Bingqin, Suvi Huikuri, Yongmei Zhang and Wenjiang Chen (2015). 'Motivating intersectoral collaboration with the Hygienic City Campaign in Jingchang, China', *Environment & Urbanization* 27:1, pp. 285-302

Li, Hongmei (2010). 'From *Chengfen* to *Shenjia* – Branding and Promotional Culture in China', in *Blowing up the Brand – Critical Perspectives on Promotional Culture*, ed. by Melissa Aronczyk and Devon Powers (New York: Peter Lang Publishing Inc.), pp. 145-169

Li, Judy (2015). 'Ways Forward from China's Urban Waste Problem', *The Nature of Cities* (http://www.thenatureofcities.com/2015/02/01/ways-forward-from-chinas-urban-waste-problem/), accessed 13 November 2015

Li, Judy, and Xie Pengfei (2015). 'Urban China's Appetite for Land', *The Nature of Cities* (http://www.thenatureofcities.com/2015/05/20/urban-chinas-appetite-for-land/), accessed 13 November 2015

Li, Shichao (2002). 'Junk-buyers as the linkage between waste sources and redemption depots in urban China: the case of Wuhan', *Resources, Conservation and Recycling* 36, pp. 319-335

Li, Wanxin, Jieyan Liu, and Duoduo Li (2012). 'Getting their voices heard: Three cases of public participation in environmental protection in China', *Journal of Environmental Management* 98, pp. 65-72

Li, Yun, Xingang Zhao, Yanbin Li, Xiaoyu Li (2015). 'Waste incineration industry and development policies in China', *Waste Management* 46, pp. 234-241

Liang, Xiaosheng (2014). *An Analysis of Chinese social classes* (中国社会各阶层分析) (Beijing: Wenhua meishu chubanshe)

Lianpeng (2017). 'If I were guiding the "garbage culture day trip"' ('假如我是"垃圾文明一日游"讲解员), *Friends of Nature* 5 May 2017 (http://www.fon.org.cn/index.php?option=com_k2&view=item&id=66...), accessed 28 August 2017

Lieberthal, Kenneth (1995). *Governing China – From Revolution Through Reform* (New York: W.W. Norton & Company Inc.)

Lin, Yi (2011). 'Turning Rurality into Modernity: Suzhi Education in a Suburban Public School of Migrant Children in Xiamen', *The China Quarterly* (2011), pp. 313-330

Lindner, Christoph, and Miriam Meissner (2016). 'Globalization, garbage, and the urban environment', in *Global Garbage – Urban imaginaries of waste, excess, and abandonment*, ed. by Christoph Lindner and Miriam Meissner (Oxon: Routledge), pp. 1-13

Linzner, Roland and Stefan Salhofer (2014). 'Municipal solid waste recycling and the significance of informal sector in urban China', *Waste Management & Research* 32:9, pp. 896-907

Liu, Ke (2017). 'An experimental APP makes appointments at home to recycle goods' (废品回收试点APP预约上门), *Beijing Ribao* 11 April 2017 (http://www.bjmac.gov.cn/zwxx/zwdtxx/mtbd/201704/t20170411_39701.html)

Liu, Lingxuan, Bing Zhang, and Jun Bi (2012). 'Reforming China's multi-level environmental governance: Lessons from the 11th Five-Year Plan', *Environmental Science and Policy* 21, pp. 106-111

Liu, Nicole Ning, Carlos Wing-Hung Lo, Xueyong Zhan, and Wei Wang (2015). 'Campaign-Style Enforcement and Regulatory Compliance', *Public Administration Review*, pp. 85-95

Liu, Ran, Tai-Chee Wong and Shenghe Liu (2012). 'Peasants' counterplots against the state monopoly of the rural urbanization process: urban villages and "small property housing" in Beijing, China', *Environment and Planning* A 44, pp. 1219-1240

Liu, Ran, Tai-Chee Wong and Shenghe Liu (2013). 'Low-wage migrants in northwestern Beijing, China: The hikers in the urbanisation and growth process', *Asia Pacific Viewpoint* 54:3, pp. 352-371

Liu, Sha (2015). 'To burn or not to burn', *The World of Chinese*, 5 October 2015 (http://www. theworldofchinese.com/2015/10/to-burn-or-not-to-burn/), accessed 6 October 2015

Liu, Yiqian (2016). *A Political Economy Analysis of the Chinese Delivery Worker's Daily Communication Practice*, Unpublished MA Thesis, Simon Fraser University

Liu, Yong, Chenjunyuan Sun, Bo Xia, Caiyun Cui, Vaughn Coffey (2018). 'Impact of community engagement on public acceptance towards waste-to-energy incineration projects: Empirical evidence from China', *Waste Management* 76, pp. 431-442

Liu, Yuting, and Fulong Wu (2006). 'The State, Institutional Transition and the Creation of New Urban Poverty in China', *Social Policy & Administration* 40:2, pp. 121-137

Lo, Kevin (2015). 'How authoritarian is the environmental governance of China', *Environmental Science & Policy* 54, pp. 152-159

Loyalka, Michelle Dammon (2012). *Eating Bitterness – Stories from the front lines of China's great urban migration* (Berkeley: University of California Press)

Lu, Hanchao (1999). 'Becoming Urban: Mendicancy and Vagrants in Modern Shanghai', *Journal of Social History* 33:1, pp. 7-36

Lu, Hongyong (2017). 'Why Shared Umbrellas are hardly taking China by Storm', *Sixth Tone* 5 June 2017, accessed 6 June 2017 (http://www.sixthtone.com/news/1000287/Why%20 Shared%20Umbrellas%20Are%20Hardly%20Taking%20China%20By%20Storm)

Lu, Jia-Wei, Sukun Zhang, Jing Hai, Ming Lei (2017). 'Status and perspectives of municipal solid waste incineration in China: A comparison with developed regions', *Waste Management* 69, pp. 170-186

Lu, Sheldon (2017). 'Introduction: Chinese-language ecocinema', *Journal of Chinese Cinemas* 2017, pp. 1-12

Lu, Yiyi (2007). 'The Autonomy of Chinese NGOs: A New Perspective', *CHINA: An International Journal* 5:2, pp. 173-203

Lucas, Gavin (2002). 'Disposability and Dispossession in the Twentieth Century', *Journal of Material Culture* 7:1, pp. 5-22

Lynteris, Christos (2013). *The Spirit of Selflessness in Maoist China: Socialist Medicine and the New Man* (Houndmills: Palgrave Macmillan)

Ma, Boyang (2017). 'Why China Is Bursting at the Seams With Discarded Clothes', *Sixth Tone* 31 August 2017 (http://www.sixthtone.com/news/1000777/why-china-is-bursting-at-the-seams-with-discarded-clothes), accessed 31 August 2017

Ma, Tianjie (2016). 'Pan Yue's vision of green China', *China Dialogue* 8 March 2016 (www.chinadialogue.net/article/show/single/en/8695-Pan-Yue-s-vision-of-green-China), accessed 10 March 2016

MacFarquhar, Roderick (1987). *The Origins of the Cultural Revolution, Volume II: The Great Leap Forward, 1958-1960* (New York: Columbia University Press)

MacFarquhar, Roderick, and Michael Schoenhals (2006). *Mao's Last Revolution* (Cambridge: The Belknap Press of Harvard University Press)

Machotka, Ewa, and Katarzyna J. Cwiertka (2016). *Too Pretty to Throw Away: Packaging Design from Japan* (Kraków: Manggha Museum of Japanese Art and Technology)

Mao, Da (2011). 'Restricted Use of Plastic Shopping Bags: The Way Out', in *The China Environment Yearbook, Volume 5, State of Change: Environmental Governance and Citizens' Rights*, ed. by Yang Dongping and Friends of Nature (Leiden, etc.: Koninklijke Brill NV), pp. 215-223

Marinelli, Maurizio (2018). 'How to Build a "Beautiful China" in the Anthropocene. The Political Discourse and the Intellectual Debate on Ecological Civilization', *Journal of Chinese Political Science* 23, pp. 365-386

Martens, Susan (2006). 'Public participation with Chinese characteristics: Citizen consumers in China's environmental management', *Environmental Politics* 15:02, pp. 211-230

McCollough, John (2007). 'The effect of income growth of the mix of purchases between disposable goods and reusable goods', *International Journal of Consumer Studies* 31, pp. 213-219

McCollough, John (2012). 'Determinants of a throwaway society – A sustainable consumption issue', *The Journal of Socio-Economics* 41, pp. 110-117

McDowall, Will, Yong Geng, Beijia Huang, Eva Barteková, Raimund Bleischwitz, Serdar Türkeli, René Kemp, and Teresa Doménech (2017). 'Circular Economy Policies in China and Europe', *Journal of Industrial Ecology* 21:3, pp. 651-661

McIsaac, Lee (2000). 'The City as Nation: Creating a Wartime Capital in Chongqing', in *Remaking the Chinese City – Modernity and National Identity, 1900-1950*, ed. by Joseph Esherick (Honolulu: University of Hawai'i Press), pp. 174-191

McKinsey & Company and Ocean Conservancy (2015). *Stemming the Tide: Land-based strategies for a plastic-free ocean, 2015* (https://oceanconservancy.org/wp-content/uploads/2017/04/full-report-stemming-the.pdf), accessed 24 November 2016

Medina, Martin (2011). 'Global supply chains in Chinese industrialization: Impact on waste scavenging in developing countries', *Working paper No. 78*, World Institute for Development Economics Research

Medina, Martin (2015). 'Living off Trash in Latin America – Debunking the Myths, *ReVista* 14:2, pp. 20-23

Melosi, Martin V. (1996). 'The Viability of Incineration as a Disposal Option – The Evolution of a Niche Technology, 1885-1995', *Public Works Management & Policy* 1:1, pp. 31-42

Meng, Dengke (2010). 'The fire-starters', *China Dialogue* 14 May 2010 (https://www.chinadialogue.net/article/show/single/en/3619-The-fire-starters), accessed 20 March 2017

MEE (Ministry of Ecology and Environment) (2018). 'MEE holds the first ministerial executive meeting', 29 March 2018 (http://english.mep.gov.cn/News_service/news_release/201803/t20180329_433300.shtml), accessed 1 May 2018

Ministry of Foreign Affairs of the People's Republic of China (2016). 'Keynote Speech by H.E. Xi Jinping, President of the People's Republic of China, at the Opening Ceremony of the B20 Summit, Hangzhou 3 September 2016', (http://www.fmprc.gov.cn/mfa_eng/wjdt_665385/zyjh_665391/t1396112.shtml), accessed 9 September 2017

Minter, Adam (2013a). *Junkyard Planet – Travels in the billion-dollar trash trade* (New York: Bloomsbury Press)

Minter, Adam (2013b). 'Junkyard Planet – A Lecture and Q&A with Adam Minter', 3 December 2013, University of Southern California Dornsife, East Asian Studies Center (http://capture.usc.edu/Mediasite/Play/72eadfbcd78543618c3281eae4d9d4b41d?catalog=69c1b544-38f3-478a-bc2a-cde59aa58b65), accessed 20 April 2016

Minter, Adam (2014). 'China's Trash Is Getting Dirtier', *Bloomberg* View 18 September 2014 (http://www.bloombergview.com/articles/2014-09-18/china-s-trash-is-getting-dirtier), accessed 12 November 2015

Minter, Adam (2015). 'The junkman is your green future', *TEDxBeijing*, June 2015 (http://v.youku.com/v_show/id_XMTM2MTM1NDYoMA==.html), accessed 26 October 2015

Mitter, Rana (2004). *A Bitter Revolution – China's Struggle with the Modern World* (Oxford: Oxford University Press)

Mobrand, Erik (2006). 'Politics of cityward migration: an overview of China in comparative perspective', *Habitat International* 30, pp. 261-274

MOHURD (Ministry of Housing and Urban-Rural Development), National Development and Reform Commission, Ministry of Land and Resources, and others (2016). 'Suggestions on Continuously Strengthening Urban Household Garbage Incineration by Ministry of Housing and Urban-Rural Development, National Development and Reform Commission, Ministry of Land and Resources, and others, Number: 227, Date of publication 22 October 2016, Date of enforcement 22 October 2016' (住房城乡建设部等部门关于进一步加强城市生活垃圾焚烧处理工作的意见) (http://www.mohurd.gov.cn/wjfb/201611/t20161105_229408.html), accessed 7 September 2017

Moore, Sarah A. (2012). 'Garbage matters: Concepts in new geographies of waste', *Progress in Human Geography* 36:6, pp. 780-799

Munro, Donald J. (1977). *The Concept of Man in Contemporary China. Michigan Studies on China* (Michigan: The University of Michigan Press)

Muramatsu, Y. (1955). 'Perspectives of the Industrialization Policy in Communist China', in *Contemporary China*, ed. by E. Stuart Kirby (London: Hong Kong University Press), pp. 83-91

Murphy, Rachel (2004). 'Turning Peasants into Modern Chinese Citizens: "Population Quality" Discourse, Demographic Transition and Primary Education', *The China Quarterly* (2004), pp. 1-20

Murray, Alan, Keith Skene, Kathryn Haynes (2017). 'The Circular Economy: An Interdisciplinary Exploration of the Concept and Application in a Global Context', *Journal of Business Ethics* 140, pp. 369-380

National Development and Reform Commission (NDRC) and Ministry of Finance (2016). 'Notification on Publishing and Distributing the Experience from Trial Spots for Developing Circular Economy by National Development and Reform Commission and Ministry of Finance, Number: 965, Date of publication 4 May 2016, Date of enforcement 4 May 2016' (国家发展改革委、财政部关于印发国家循环经济试点示范典型经验的通知) (http://www.ndrc.gov.cn/zcfb/zcfbtz/201605/t20160510_801123.html), accessed 1 November 2017

National People's Congress Standing Committee (2008). 'Circular Economy Promotion Law of the People's Republic of China, issued on 29 August 2008, effective as of 1 January 2009' (http://www.lawinfochina.com/display.aspx?id=7025&lib=law), accessed 1 March 2018

Naustdalslid, Jon (2014). 'Circular economy in China – the environmental dimension of the harmonious society', *International Journal of Sustainable Development & World Ecology* 21:4, pp. 303-313

Nelles, Michael, et al. (2017). *Recycling and Recovery of the biogenic fractions from municipal solid waste in the PR of China* (Rostock, Universität Rostock)

Ngeow, Chow Bing (2012). 'The Residents' Committee in China's Political System: Democracy, Stability, Mobilization', *Issues & Studies* 48:2, pp. 71-126

Norcliffe, Glen (2011). 'Neoliberal mobility and its discontents: Working tricycles in China's cities', *City, Culture and Society* 2, pp. 235-242

Nowling, Una (2016). 'Waste to Energy: An Opportunity Too Good to Waste, or a Waste of Time?', *Powermag* 1 September 2016 (http://www.powermag.com/waste-energy-opportunity-good-waste-waste-time/?printmode=1), accessed 12 March 2017

O'Brien, Martin (1999). 'Rubbish values: Reflections on the political economy of waste', *Science as Culture* 8:3, pp. 269-295

O'Brien, Martin (2013). 'Consumers, Waste and the "Throwaway Society" Thesis: Some Observations on the Evidence', *International Journal of Applied Sociology* 3:2, pp. 19-27

Office of the Capital Spiritual Civilization Construction Committee, Beijing Municipal Appearance Management Committee (eds.) (2010a). *Green Frog in Action – A Pictorial Guide to Garbage Reduction and Garbage Classification* (绿娃在行动 – 图说垃圾减量垃圾分类) (Beijing: Beijing chubanshe)

Office of the Capital Spiritual Civilization Construction Committee, Beijing Municipal Appearance Management Committee (eds.) (2010b). *Household Waste Classification Guidebook for Citizens of the Capital* (首都市民生活垃圾分类指导手册) (Beijing: n.p.)

Office of the Capital Spiritual Civilization Construction Committee, Beijing Municipal Appearance Management Committee (2010c). '"Be a Polite Beijinger, Reduce the Garbage Amount and Garbage Classification starting from Oneself" Plan' (关于印发《"做文明有礼的北京人，垃圾减量垃圾分类从我做起"主题宣传实践活动方案》的通知), [2010]7 (http://zt.bjwmb.gov.cn/ljjlfl/xgwj/)

Office of the Capital Spiritual Civilization Construction Committee, Beijing Municipal Appearance Management Committee, Beijing Municipal Education Committee (2014). *The Story of Garbage* (垃圾的故事) (Beijing: Beijing chubanshe)

Ou, Tzu-Chi (2011). 'Hiding a "Garbage Village": Changes in Urban Governance at the 2008 Beijing Olympics', AACS conference 2 October 2011

Pasquier, Martin (2015). 'Internet Plus: China's official strategy for the uberisation of the economy', *Innovation is Everywhere* 1 May 2015 (http://www.innovationiseverywhere.com/internet-plus-chinas-official-strategy-for-the-uberisation-of-the-economy/), accessed 3 January 2017

PKU MBA Deep Dive (2015). 'The Problem with Recycling in Beijing', *The Beijinger* 27 November 2015 (http://www.thebeijinger.com/blog/2015/11/27/pku-mba-deep-dive-problem-recycling-beijing), accessed 6 March 2017

Poole, Steven (2018). 'Plogging: the fitness craze that's sweeping the streets', *The Guardian* 31 March 2018 (https://www.theguardian.com/books/2018/mar/31/plogging-steven-poole), accessed 12 April 2018

Prasad, Sameer, Ashish Jain, Jasmine Tata, and Shantha Parthan (2012). 'From Rags to Riches: Tapping the Social Capital within the Solid Waste Informal Sector', *South Asian Journal of Business and Management Cases* 1:2, pp. 77-89

PressTV Reporter's File (2014). 'Chinese vending machines pay in cash for garbage', *PressTV News Videos* 28 November 2014 (http://217.218.67.229/detail/2014/11/28/387863/recycling-vending-machines-in-china/), accessed 22 August 2016

Qi, Jianguo, Bin Wu, Wen-Jun Li, Hong Wang, Jingxing Zhao, and Xushu Peng (2016). *Development of Circular Economy in China* (Singapore: Springer)

Qu, Xiao-yan, Zhen-shan Li, Xin-yuan Xie, Yu-mei Sui, Lei Yang, and You Chen (2009). 'Survey of composition and generation rate of household wastes in Beijing, China', *Waste Management* 29, pp. 2618-2624

Ran, Ran (2013). 'Perverse Incentive Structure and Policy Implementation Gap in China's Local Environmental Politics', *Journal of Environmental Policy & Planning* 15:1, pp. 17-39

Reno, Joshua Ozias (2014). 'Toward a New Theory of Waste: From "Matter out of Place" to Signs of Life', *Theory, Culture & Society* 31:6, pp. 3-27

Rogaski, Ruth (2000). 'Hygienic Modernity in Tianjin', in *Remaking the Chinese City – Modernity and National Identity, 1900-1950*, ed. by Joseph Esherick (Honolulu: University of Hawai'i Press), pp. 30-46

Russo, Alessandro (2012). 'How Did the Cultural Revolution End? The Last Dispute between Mao Zedong and Deng Xiaoping, 1975', *Modern China* 39:3, pp. 239-279

Saich, Tony (2000). 'Negotiating the State: The Development of Social Organizations in China', *The China Quarterly* (2000), pp. 124-141

Saich, Tony (2008). 'The Changing Role of Urban Government', in *China Urbanizes: Consequences, Strategies, and Policies*, ed. by Shahid Yusuf and Tony Saich (Washington, D.C.: World Bank), pp. 181-206

Saich, Tony (2016). 'How China's citizens view the quality of governance under Xi Jinping', *Journal of Chinese Governance* 1:1, pp. 1-20

Salmenkari, Taru (2008). 'Searching for a Chinese Civil Society Model', *China Information* 23:3, pp. 397-421

Saunders, Peter, and Sun Lujun (2006). 'Poverty and Hardship among the Aged in Urban China', *Social Policy & Administration* 40:2, pp. 138-157

Schmitz, Rob (2017). 'The Burning Problem of China's Garbage', *National Public Radio* 20 February 2017 (https://www.npr.org/sections/parallels/2017/02/20/515814016/the-burning-problem-of-chinas-garbage), accessed 10 March 2017

Schwartz, Jonathan (2004). 'Environmental NGOs in China: Roles and Limits', *Pacific Affairs* 77:1, pp. 28-49

Shapiro, Judith (2001). *Mao's War Against Nature – Politics and the Environment in Revolutionary China* (Cambridge: Cambridge University Press)

Shao, Liming, He Pinjing, and Liu Yongde (2007). 'Factors Affecting the Separation Quality of Source-Separated Collection for Rural Waste', *Journal of Agro-Environment Science* 26:1, pp. 326-329

Sheridan, James E. (1975). *China in Disintegration: The Republican Era in Chinese History, 1912-1949* (New York: The Free Press)

Shi, Rui, Xu Heqian, Zhang Boling, and Yao Jiayi (2015). 'Slowly, China Prepares to Raise Retirement Age', *Caixin Online* 1 April 2015 (http://english.caixin.com/2015-04-01/100796734.html), accessed 27 August 2015

Shi-Kupfer, Kristin, Mareike Ohlberg, Simon Lang, and Bertram Lang (2017). 'Ideas and Ideologies Competing For China's Political Future – How online pluralism challenges official orthodoxy', *MERICS Papers on China No. 5*, October 2017

Shin, Hyun Bang (2009). 'Residential Redevelopment and the Entrepreneurial Local State: The Implications of Beijing's Shifting Emphasis on Urban Redevelopment Policies', *Urban Studies* 46:13, pp. 2815-2839

Shin, Hyun Bang, and Bingqin Li (2013). 'Whose games? The costs of being "Olympic citizens" in Beijing', *Environment & Urbanization* 25:2, pp. 559-576

Sigley, Gary (2009). '*Suzhi*, the Body, and the Fortunes of Technoscientific Reasoning in Contemporary China', *positions: east asia cultures critique* 17:3, pp. 537-566

Simões, Fernando Dias (2016). 'Consumer Behavior and Sustainable Development in China: The Role of Behavioral Sciences in Environmental Policymaking', *Sustainability* 8:897

Solinger, Dorothy J. (1999). *Contesting Citizenship in Urban China – Peasant Migrants, the State, and the Logic of the Market* (Berkeley: University of California Press)

Solinger, Dorothy J. (2004). 'The new crowd of the dispossessed – The shift of the urban proletariat from master to mendicant', in *State and Society in 21st-century China – Crisis, contention, and legitimation*, ed. by Peter Hays Gries and Stanley Rosen (New York: RoutledgeCurzon), pp. 50-66

Solinger, Dorothy J. (2006). 'The creation of a new underclass in China and its implications', *Environment & Urbanization* 18:1, pp. 177-193

Someno, Kenji (2014). 'Recycling and Economic Growth in China's Interior', *The Tokyo Foundation* (http://www.tokyofoundation.org/en/articles/2014/recycling-in-china-interior), accessed 5 June 2016

Spires, Anthony J. (2011). 'Contingent Symbiosis and Civil Society in an Authoritarian State: Understanding the Survival of China's Grassroots NGOs', *American Journal of Sociology* 117:1, pp. 1-45

State Council (2014). 'Notice concerning Issuance of the Planning Outline for the Construction of a Social Credit System (2014-2020) GF No. (2014)21, 14 June 2014', *The State Council of the People's Republic of China, 14 June 2014* (https://chinacopyrightandmedia.wordpress.com/2014/06/14/planning-outline-for-the-construction-of-a-social-credit-system-2014-2020/), accessed 23 April 2018

State Council (2015a). 'Report on the Work of the Government, 16 March 2015', *The State Council of the People's Republic of China* (http://english.gov.cn/archive/publications/2015/03/05/content_281475066179954.htm), accessed 8 January 2017

State Council (2015b). 'Instructions from the State Council to actively promote "Internet+"' (国务院关于积极推进"互联网+"行动的指导意见), *The State Council of the People's Republic of China*, 4 July 2015 (http://www.gov.cn/zhengce/content/2015-07/04/content_10002.htm), accessed 8 January 2017

State Council (2015c). 'Made in China 2025, Number [2015]28, issued 8 May 2015, in force from 19 May 2015', *The State Council of the People's Republic of China*, 8 May 2015 (http://www.gov.cn/zhengce/content/2015-05/19/content_9784.htm), accessed 10 November 2018

State Council (2017). 'Action plan to phase out waste imports', *The State Council of the People's Republic of China*, 27 July 2017 (http://english.gov.cn/policies/latest_releases/2017/07/27/content_281475756814340.htm), accessed 1 November 2017

State Council General Office (2017). 'Daily Garbage Classification System Plan drawn up by the Ministry of Housing and Urban-Rural Development and the National Development and Reform Commission, Number [2017]26, issued 18 March 2017, in force from 18 March 2017' (国家发展改革委、住房城乡建设部生活垃圾分类制度实施方案), *The State Council of the People's Republic of China*, 18 March 2017 (http://www.gov.cn/zhengce/content/2017-03/30/content_5182124.htm), accessed 1 November 2017

State Grid Corporation of China (2016). 'Beijing Gao Antun Electric Vehicle Recharging & Switching Station Officially Put into Operation', *State Grid News* (http://www.sgcc.com.cn/ywlm/mediacenter/corporatenews/03/269446.shtml), 16 March 2016, accessed 1 November 2017

Steinhardt, H. Christoph (2012). 'How is High Trust in China Possible? Comparing the Origins of Generalized Trust in Three Chinese Societies', *Political Studies* 60, pp. 434-454

Steinhardt, H. Christoph, and Fengshi Wu (2015). 'In the Name of the Public: Environmental Protest and the Changing Landscape of Popular Contention in China', *The China Journal* 75, pp. 61-82

Steuer, Benjamin, Roland Ramusch, Florian Part, and Stefan Salhofer (2017). 'Analysis of the value chain and network structure of informal waste recycling in Beijing, China', *Resources, Conservation and Recycling* 117, pp. 137-150

Strand, David (1989). *Rickshaw Beijing – City People and Politics in the 1920s* (Berkeley: University of California Press)

Strasser, Susan (2003). 'The Alien Past: Consumer Culture in Historical Perspective', *Journal of Consumer Policy* 26, pp. 375-393

Strauss, Julia (2006). 'Morality, Coercion and State Building by Campaign in the Early PRC: Regime Consolidation and After, 1949-1956', *The China Quarterly* (2006), pp. 891-912

Su, Yuting (2014). 'Beijing trials bottle-recycling vending machines', *China Central Television* 21 September 2014 (http://english.cntv.cn/2014/09/21/VIDE1411276563132182.shtml), accessed 29 September 2014

Sun, Ivan Y., Rong Hu, Daniel F.K. Wong, Xuesong He, and Jessica C.M. Li (2013). 'One country, three populations: Trust in police among migrants, villagers, and urbanites in China', *Social Science Research* 42, pp. 1737-1749

Sun, Wanning (2009). '*Suzhi* on the Move: Body, Place, and Power', *positions: east asia cultures critique* 17:3, pp. 617-642

Sun, Wenkai, and Xianghong Wang (2012). 'Do government actions affect social trust? Cross-city evidence in China', *The Social Science Journal* 49, pp. 447-457

Swiss RE (2016). 'Analysis of Tianjin Port Explosion: Risk management is the key', *Swiss RE* (http://www.swissre.com/china/Analysis_of_Tianjin_Port_Explosion.html), June 2016, accessed 1 March 2017

Tai, Jun, Weiqian Zhang, Yue Che, and Di Feng (2011). 'Municipal solid waste source-separated collection in China: A comparative analysis', *Waste Management* 31, pp. 1673-1682

Tan, Soo Jiuan, and Siok Kuan Tambyah (2011). 'Generalized Trust and Trust in Institutions in Confucian Asia', *Social Indicators Research* 103:3, pp. 357-377

Tang, Wenfang (2004). 'Interpersonal Trust and Democracy in China, Conference Paper Presented at the International Conference "The Transformation of Citizen Politics and Civi Attitudes in Three Chinese Societies"', 2004

Tang, Wenfang (2018). 'The "Surprise" of Authoritarian Resilience in China', *American Affairs* 2:1 (https://americanaffairsjournal.org/2018/02/surprise-authoritarian-resilience-china/), accessed 5 April 2018

Tang, Zhongjun, Xiaohong Chen, and Jianghong Luo (2011). 'Determining Socio-Psychological Drivers for Rural Household Recycling Behavior in Developing Countries: A Case Study from Wugan, Hunan, China', *Environment and Behavior* 43:6, pp. 848-877

Taylor, Alan (2018). 'The Bike-Share Oversupply in China: Huge Piles of Abandoned and Broken Bicycles', *The Atlantic* 22 March 2018 (https://www.theatlantic.com/photo/2018/03/bike-share-oversupply-in-china-huge-piles-of-abandoned-and-broken-bicycles/556268/), accessed 28 March 2018

Taylor, John G. (2008). 'Poverty and Vulnerability', in *China Urbanizes: Consequences, Strategies, and Policies*, ed. by Shahid Yusuf and Tony Saich (Washington, D.C.: World Bank), pp. 91-104

Teets, Jessica C. (2013). 'Let Many Civil Societies Bloom: The Rise of Consultative Authoritarianism in China'. *The China Quarterly* (2013), pp. 1-20

Teets, Jessica C. (2014). 'Civil Society in China'. In *Civil Society under Authoritarianism: The China Model* (Cambridge: Cambridge University Press), pp. 1-37

Tian, Youyi, and Chenyu Wang (2016). 'Environmental Education in China: Development, Difficulties and Recommendations, *Journal of Social Science Studies*, 3:1, pp. 31-43

Tomba, Luigi (2009). 'Of Quality, Harmony, and Community: Civilization and the Middle Class in Urban China', *positions: east asia cultures critique* 17:3, pp. 592-616

Tong, Xin (2017). 'Waste is "wicked" when we try to solve it. Author's response to Joshua Goldstein's comments', *Resources, Conservation and Recycling* 117, pp. 175-176

Tong, Xin, and Dongyan Tao (2016). 'The rise and fall of a "waste city" in the construction of an "urban circular economic system": The changing landscape of waste in Beijing', *Resources, Conservation and Recycling* 107, pp. 10-17

Tsang, Eileen Yuk-ha, and Pak K. Lee (2013). 'The Chinese New Middle Class and Green NGOs in South China: Vanguards of Guanxi (Connections)-Seeking, Laggards in Promoting Social Causes?', *China: An International Journal* 11:2, pp. 155-169

Tse, Chun Wing (2016). 'Urban Residents' Prejudice and Integration of Rural Migrants into Urban China', *Journal of Contemporary China* 25:100, pp. 579-595

Tu, Qin, Arthur P.J. Mol, Lei Zhang, Ruerd Ruben (2011). 'How do trust and property security influence household contributions to public goods? The case of the sloping land conversion program in China', *China Economic Review* 22, pp. 499-511

United Nations Framework Convention on Climate Change (no date). *The Paris Agreement* (http://unfccc.int/paris_agreement/items/9485.php), last accessed 1 March 2018

Upton-McLaughlin, Sean (2014). 'The many faces of suzhi in the Chinese organization and society: Implications for multinational HRM practice', *Journal of Chinese Human Resources Management* 5:1, pp. 51-61

Van Dam, Peter, and Joost Jonker (2017). 'Introduction – The Rise of Consumer Society', *Low Countries Historical Review*, 132:3, pp. 3-10

Van Rooij, Benjamin (2010). 'The People vs. Pollution: understanding citizen action against pollution in China', *Journal of Contemporary China* 19:63, pp. 55-77

Van Rooij, Benjamin (2012). 'The People's Regulation: Citizens and Implementation of Law in China', *Columbia Journal of Asian Law* 25:2, pp. 116-179

Van Rooij, Benjamin, Adam Fine, Yanyan Zhang, and Yunmei Wu (2017). 'Comparative Compliance: Digital Piracy, Deterrence, Social Norms and Duty in China and the United States', *Law & Policy* 39:1, pp. 73-93

Vanacore, Tara Sun (2012a). 'Refusing to Waste Away: China's Tale of Trash Cities and the Incinerator Boom, Part I Snapshot of China's Waste Challenge', *China Environment Forum, Wilson Center*, November 2012

Vanacore, Tara Sun (2012b). 'Wasting No Time: A Chinese NGO's Campaign for Waste Management Activism, Part II Snapshot of China's Waste Challenge', *China Environment Forum, Wilson Center*, November 2012

Varul, Matthias Zick (2006). 'Waste, Industry and Romantic Leisure – Veblen's Theory of Recognition', *European Journal of Social Theory* 9:1, pp. 103-117

Visser, Carolijn (2016). *Selma – Aan Hitler ontsnapt, gevangene van Mao* (Amsterdam: Atlas Contact)

Vlahov, David, Nicholas Freudenberg, Fernando Proietti, Danielle Ompad, Andrew Quinn, Vijay Nandi, and Sandro Galea (2007). 'Urban as a Determinant of Health', *Journal of Urban Health* 84:1, pp. 116-126

Wallace, Jeremy L. (2014). 'Juking the Stats? Authoritarian Information Problems in China', *British Journal of Political Science* 46, pp. 11-29

Wallenwein, Fabienne (2013). *The Housing Model xiaoqu 小区: the Expression of an Increasing Polarization of the Urban Population in Chinese Cities?*, Unpublished master's thesis Heidelberg University, 9 December 2013

Walsh, Matthew (2017). 'Let's Make This Year's Singles' Day the Last One Ever', *Sixth Tone* 11 November 2017 (http://www.sixthtone.com/news/1001163/lets-make-this-years-singles-day-the-last-one-ever), accessed 11 November 2017

Wan, Zheng, Jihong Chen, Brian Craig (2015). 'Lessons learned from Huizhou, China's unsuccessful waste-to-energy incinerator project: Assessment and policy recommendations', *Utilities Policy* 33, pp. 63-68

Wang, Di (2013). 'Operating Norms and Practices of Residents' Committees – The consequences and limits of management by numbers', *China Perspectives* 1 (2013), pp. 7-15

Wang, Hua, Jie He, Yoonhee Kim, and Takuya Kamata (2011). 'Municipal Solid Waste Management in Small Towns – An Economic Analysis Conducted in Yunnan, China', *Policy Research*

Working Paper 5767, The World Bank Development Research Group Environment and Energy Team, August 2011

Wang, Jia, Ling Han, and Shushu Li (2008). 'The collection system for residential recyclables in communities in Haidian District, Beijing: A possible approach for China recycling', *Waste Management* 28, pp. 1672-1680

Wang, Jiuliang (王久良) (2011a). *Beijing Besieged by Waste* (垃圾围城), Icarus Films 2011 (https://www.youtube.com/watch?v=W73eKAjyNXs, published on 8 November 2013), accessed 16 September 2015

Wang, Jiuliang (2011b). 'Beijing Besieged by Garbage', *Cross-Currents: East Asian History and Culture Review* 1 (https://cross-currents.berkeley.edu/e-journal/photo-essay/beijing-besieged-garbage/statement), accessed 12 November 2015

Wang, Meiqin (2017). 'Waste in contemporary Chinese Art', *IIAS Newsletter* 76, pp. 32-33

Wang, Xiaoying (2002). 'The Post-Communist Personality: The Spectre of China's Capitalist Market Reforms', *The China Journal* 47, pp. 1-17

Wang, Xinhong (2016). 'Requests for Environmental Information Disclosure in China: an understanding from legal mobilization and citizen activism', *Journal of Contemporary China* 25:98, pp. 233-247

Wang, Yu, Xu Shi-wei, Yu Wen, Ahmed Abdul-gafar, Liu Xiao-jie, Bai Jun-fei, Zhang Dan, Gao Li-wei, Cao Xiao-chang, and Liu Yao (2016). 'Food packing: A case study of dining out in Beijing', *Journal of Integrative Agriculture* 15:8, pp. 1924-1931

Wang, Zheng (2013). 'The Chinese Dream: Concept and Context', *Journal of Chinese Political Science* 19, pp. 1-13

'Waste sorting me first, Charming Wuxi I add green' (垃圾分类我先行 魅力无锡我添绿) (2014) (http://wx.sina.com.cn/video/interview/2014-03-20/13105089.html), 20 March 2014, accessed 10 March 2017

Watts, Jonathan (2010). 'Beijing to sweeten stench of rubbish crisis with giant deodorant guns', *The Guardian* 26 March 2010 (https://www.theguardian.com/environment/2010/mar/26/beijing-rubbish-deodorant), accessed 3 February 2017

Wei, Yuan-Song, Yao-Bo Fan, Min-Jian Wang, and Ju-Si Wang (2000). 'Composting and compost application in China', *Resources, Conservation and Recycling* 30, pp. 277-300

Williams, Adam S. (2014). *Excess and Access: Informal Recycling Networks and Participants in Shanghai, China*. PhD thesis University of Colorado 2014 *Geography Graduate Theses & Dissertations*. Paper 4

Wilson, David C., Costas Velis, and Chris Cheeseman (2006). 'Role of informal sector recycling in waste management in developing countries', *Habitat International* 30, pp. 797-808

Wong, Daniel Fu Keung, Chang Ying Li, and He Xue Song (2007). 'Rural migrant workers in urban China: living a marginalised life', *International Journal of Social Welfare* 16, pp. 32-40

Wong, Natalie W.M. (2016). 'Environmental Protests and NIMBY activism: Local politics and waste management in Beijing and Guangzhou', *China Information* 30:2, pp. 143-164

Wu, Fengshi, and Kin-Man Chan (2012). 'Graduated Control and Beyond – The Evolving Government-NGO Relations', China Perspectives 3 (2012), pp. 9-17

Wu, Hung (1991). 'Tiananmen Square: A Political History of Monuments', *Representations* 35, pp. 84-117

Wu, Hung (2005). *Remaking Beijing – Tiananmen Square and the Creation of a Political Space* (Chicago: University of Chicago Press, 2005)

Wu, Ka-ming and Zhang Jieying (2016). *Living with Waste: Economies, Communities and Spaces of Waste Collectors in China* (废品生活 – 垃圾场的经济，社群与空间) (Hong Kong: Chinese University of Hong Kong)

Wu, Xiaobo (2016). 'Six directions of the Central Economic Work Conference', *China Daily* 20 December 2016 (http://www.chinadaily.com.cn/business/2016top10/2016-12/20/content_27715917_4.htm), accessed 10 January 2017

Wu, Zuqiang (2002). 'Green Schools in China', *The Journal of Environmental Education* 34:1, pp. 21-25

Wübbeke, Jost, Mirjam Meissner, Max. J. Zenglein, Jacqueline Ives, Björn Conrad (2016). 'Made in China 2025 – The making of a high-tech superpower and consequences for industrial countries', *MERICS Papers on China* No. 2, December 2016 (https://www.merics.org/sites/default/files/2017-09/MPOC_No.2_MadeinChina2025.pdf), accessed 10 November 2018

Xie, Lei (2011). 'Environmental Justice in China's Urban Decision-Making', *Taiwan in Comparative Perspective* 3, pp. 160-179

Xie, Lei (2016). 'Environmental governance and public participation in rural China', *China Information* 30:2, pp. 188-208

Xie, Pengfei (2015). 'Turning Migrant Workers into Citizens in Urbanizing China', *The Nature of Cities* (http://www.thenatureofcities.com/2015/10/19/turning-migrant-workers-into-citizens-in-urbanizing-china/), accessed 5 September 2016

Xinhua (2016). 'China jails 49 for catastrophic Tianjin warehouse blast', *Xinhua* 9 November 2016 (http://www.xinhuanet.com/english/2016-11/09/c_135817728.htm), accessed 1 March 2017

Xing, Fangqun (1984). 'Some Viewpoints Concerning the Propagation of Advanced Personages and Models' (关于宣传先进人物和先进典型的几点意见), in *Zhongguo xinwen nianjian 1984* (Yearbook of Chinese News 1984), ed. by Zhongguo shehui kexueyuan xinwen yanjiusuo (Beijing: Renmin ribao chubanshe), pp. 124-128

Xiong, Wenhui and Sun Shuiyu (2004). 'A Study on Classification and Collection of Household Domestic Refuse in Guangzhou', *Journal of Guangdong University of Technology* 21:3, pp. 15-20

Xu, Feng (2008). 'Gated Communities and Migrant Enclaves: the conundrum for building "harmonious community/shequ"', *Journal of Contemporary China* 17:57, pp. 633-651

Xu, Suyun, Hongfu He and Liwen Luo (2016). 'Status and Prospects of Municipal Solid Waste to Energy Technologies in China', in *Recycling of Solid Waste for Biofuels and Bio-chemicals*, ed. by Obulisamy Parthiba Karthikeyan, Kirsten Heimann, and Subramanian Senthilkannan Muthu (Singapore: Springer Singapore), pp. 31-54

Xue, Bing, Yong Geng, Wanxia Ren, Zilong Zhang, Weiwei Zhang, Chenyu Lu, and Xingpeng Chen (2011). 'An overview of municipal solid waste management in Inner Mongolia Autonomous Region, China', *Journal of Material Cycles and Waste Management* 13, pp. 283-292

Yan, Huiqi, Benjamin van Rooij, Jeroen van der Heijden (2016). 'The enforcement-compliance paradox: Implementation of pesticide regulation in China', *China Information* 30:2, pp. 209-231

Yang, Changjiang (2011). '2009: A Waste Crisis at a Crossroads', in *The China Environment Yearbook, Volume 5, State of Change: Environmental Governance and Citizens' Rights*, ed. by Yang Dongping and Friends of Nature (Leiden: Koninklijke Brill NV), pp. 187-196

Yang, Changjiang (2013). 'The Waste Crisis: Seeking a New Direction in the Dilemma', in *Chinese Research Perspectives on the Environment, Volume 1: Urban Challenges, Public Participation, and Natural Disasters*, ed. by Yang Dongping (Leiden: Koninklijke Brill NV), pp. 175-188

Yang, Fan (2016). 'From Bandit Cell Phones to Branding the Nation – Three Moments of Shanzhai in WTO-era China', *positions: east asia cultures critique* 24:3, pp. 589-619

Yang, Guobin (2005). 'Environmental NGOs and Institutional Dynamics in China', *The China Quarterly* (2005), pp. 46-66

Yang, Lei, Zhen-Shan Li and Hui-Zhen Fu (2011). 'Model of Municipal Solid Waste Source Separation Activity: A Case Study of Beijing', *Journal of the Air & Waste Management Association* 61:2, pp. 157-163

Yang, Nianqun (2004). 'Disease Prevention, Social Mobilization and Spatial Politics: The Anti
    Germ-Warfare Incident of 1952 and the "Patriotic Health Campaign"', *The Chinese Historical
    Review* 11:2, pp. 155-182
Yang, Qing, and Wenfang Tang (2010). 'Exploring the Sources of Institutional Trust in China:
    Culture, Mobilization, or Performance?', *Asian Politics & Policy* 2:3, pp. 415-436
Yang, Shi, and Christine Furedy (1993). 'Recovery of Wastes for Recycling in Beijing', *Environmental
    Conservation* 20:1, pp. 79-82
Yang, Zan, and Yue Shen (2008). 'The affordability of owner occupied housing in Beijing', *Journal
    of Housing and the Built Environment* 23, pp. 317-335
Yates, Michelle (2011). 'The Human-As-Waste, the Labor Theory of Value and Disposability in
    Contemporary Capitalism', *Antipode* 43:5, pp. 1679-1695
Yew, Wei Lit (2017). 'Disembedding lawful activism in contemporary China: The confrontational
    politics of a green NGO's legal mobilization', *China Information*, pp. 1-20
Yin, Dafei and Xiaomei Tan (2017). 'Bike-sharing Data and Cities: Lessons From China's Experience,
    The Wilson Center New Security Beat 30 November 2017 (https://www.newsecuritybeat.
    org/2017/11/bike-sharing-data-cities-lessons-chinas-experience/), accessed 3 April 2018
Yuan, Suwen and Li Rongde (2017). 'Burning Unsorted Garbage Could Cost Beijing Billions in
    Health Care Costs', *Caixin Global* 24 March 2017 (http://www.caixinglobal.com/2017-03-
    24/101069583.html), accessed 25 March 2017
Yuan, Yalin, and Mitsuyasu Yabe (2014). 'Residents' Willingness to Pay for Household Kitchen
    Waste Separation Services in Haidian and Dongcheng Districts, Beijing City', *Environments*
    1, pp. 190-207
Yuan, Yalin, Hisako Nomura, Yoshifumi Takahashi, and Mitsuyasu Yabe (2016). 'Model of Chinese
    Household Kitchen Waste Separation Behavior: A Case Study in Beijing City', *Sustainability*
    1083:8
Yuan, Zengwei, Jun Bi, and Yuichi Moriguichi (2006). 'The Circular Economy: A New Development
    Strategy in China', *Journal of Industrial Ecology* 10:1-2, pp. 4-8
Yuen, Samson (2018). 'Negotiating Service Activism in China: The Impact of NGOs' Institutional
    Embeddedness in the Local State', *Journal of Contemporary China* 27:111, pp. 406-422

Zanasi, Margherita (2015). 'Frugal Modernity: Livelihood and Consumption in Republican China',
    Journal of Asian Studies 74:2, 391-409
Zhan, Xueyong, and Shui-Yan Tang (2013). 'Political Opportunities, Resource Constraints and
    Policy Advocacy of Environmental NGOs in China', *Public Administration* 91:2, pp. 381-399
Zhang, Chun (2015). 'China trials environmental audits to hold officials to account', *China
    Dialogue*, 18 June 2015 (https://www.chinadialogue.net/article/show/single/en/7990-China-
    trials-environmental-audits-to-hold-officials-to-account), accessed 17 March 2018
Zhang, Dong Qing, Soon Keat Tan, and Richard M. Gersberg (2010). 'Municipal solid waste
    management in China: Status, problems and challenges', *Journal of Environmental Manage-
    ment* 91, pp. 1623-1633
Zhang, Hua, Zong-Guo Wen (2014). 'The consumption and recycling collection system of PET
    bottles: A case study of Beijing, China', *Waste Management* 2014, pp. 987-998
Zhang, Li (2002). 'Spatiality and Urban Citizenship in Late Socialist China', *Public Culture* 14:2,
    pp. 311-334
Zhang, Li, and Meng Li (2016). 'Local Fiscal Capability and Liberalization of Urban Hukou',
    Journal of Contemporary China 25:102, pp. 893-907
Zhang, Shasha (2017). 'An ordinary position, a sacred duty – A visit with Li Xuewen, garbage
    separation instructor (平凡的岗位神圣的职责 – 记垃圾分类指导员李学文), *Municipal
    Management Science and Technology* 2016:5 (2017), pp. 72-73

Zhang, Ye (2016). 'App developers rush to grab share of estimated $100b recycling industry', *Global Times* 12 January 2016 (http://www.globaltimes.cn/content/963157.shtml), accessed 25 August 2016

Zhang, Yongmei, and Bingqin Li (2011). 'Motivating service improvement with awards and competitions – hygienic city campaigns in China', *Environment and Urbanization* 23, pp. 41-56

Zhao, Han (2016). 'Being a waste picker in Beijing for 20 years: your capital city, my waste capital city' (北京拾荒20年：你的京城，我的废都), *Duan Media* (端传媒), 14 September 2016 (https://theinitium.com/article/20160914-mainland-scavengers/), accessed 4 January 2018

Zhao, Pengjun, and Mengzhu Zhang (forthcoming). 'Informal suburbanization in Beijing: An investigation of informal gated communities on the urban fringe', *Habitat International*

Zhejiang Provincial Committee, Communist Party of China (2016). 'Lucid Waters and Lush Mountains Are Invaluable Assets', *Qiushi Journal* 8:2 (27) (http://english.qstheory.cn/2016-06/07/c_1118896002.htm), accessed 3 December 2017

Zheng, Jinran (2017). 'Seniors aid the recycling trend', *China Daily* 10 February 2017 (http://www.chinadaily.com.cn/china/2017-02/10/content_28158996.htm), accessed 20 February 2017

Zheng, Shuping (2017). 'Interview: Let the recycling of waste materials realize the new value of circular economy industry' (让垃圾再生利用实现循环经济产业新价值), *Municipal Management Science and Technology* 2017:1, pp. 60-65

Zheng, Siqi, Fenjie Long, C. Cindy Fan, and Yizhen Gu (2009). 'Urban Villages in China: A 2008 Survey of Migrant Settlements in Beijing', *Eurasian Geography and Economics* 50:4, pp. 425-446

Zhong, Yang (2014). 'Do Chinese People Trust Their Local Government, and Why?', *Problems of Post-Communism* 61:3, pp. 31-44

Zhong, Yang, and Wonjae Hwang (2016). 'Pollution, Institutions and Street Protests in Urban China', *Journal of Contemporary China* 25:98, pp. 216-232

Zhu, Jinxin (2013). *A Comparative Study of Qingdao, China and Hamilton, Canada from a Municipal Solid Waste Management System Perspective*, McMaster University School of Engineering Practice, *MEPP Inquiry 2013* (unpaginated)

Zhu, Minghua, Fan Xiumin, Alberto Rovetta, He Qichang, Federico Vicentini, Liu Bingkai, Alessandro Giusti, and Liu Yi (2009). 'Municipal solid waste management in Pudong New Area, China', *Waste Management* 29, pp. 1227-1233

Zhu, Xiao, Lei Zhang, Ran Ran, and Arthur P.J. Mol (2015). 'Regional restrictions on environmental impact assessment approval in China: the legitimacy of environmental authoritarianism', *Journal of Cleaner Production* 92, pp. 100-108

Zhu, Zi (2017). 'Backfired Government Action and the Spillover Effect of Contention: A Case Study of the Anti-PX Protests in Maoming, China', *Journal of Contemporary China* 26:106, pp. 521-535

Zhuang, Zhengyue, Wu Xuefei, and Jia Yue (2018). 'Hangzhou Receives First National Demonstration City Award for the Construction of the Social Credit System' (杭州市荣获全国首批社会信用体系建设示范城市), *Zhejiang Cities and Towns Web* [浙江城镇网], 10 January 2018 (http://town.zjol.com.cn/cstts/201801/t20180110_6292357.shtml), accessed 18 April 2018

Websites

AlaHB (Shanghai) http://www.alahb.com/IndexNew.aspx, accessed 6 March 2018

Bangdaojia http://www.365bdj.com/, accessed 6 March 2018

Beijing Incom Resources Recovery Recycling http://www.incomrecycle.com/, accessed 6 March 2018

Beijing Municipal Commission of City Management http://www.bjmac.gov.cn/hjwsbz/, accessed 6 March 2018

Beijing TV (BTV) Station Life Channel https://www.youtube.com/channel/UCaMJ1qzUD-gev9N1ye_uKJoA, accessed 6 March 2018

Cambridge Dictionary (Uberization) https://dictionary.cambridge.org/dictionary/english/uberize, accessed 6 March 2018

Garbage Matters: a Comparative History of Waste in East Asia (NWO 277-53-006) https://www.garbagemattersproject.com/, accessed 6 March 2018

Huishouge (Recycle Brother) http://www.huishouge.cn/, accessed 19 November 2018

Incom vacancies http://www.incomrecycle.com/recruit/, accessed 6 March 2018

Leading Group for the Implementation of the Propaganda Movement 'Be a civilized and polite Beijinger – waste reduction, garbage classification starts with me' under the Office for the Creation of the Capital's Spiritual Civilization Committee and the Beijing Municipal Appearance Management Committee http://zt.bjwmb.gov.cn/ljjlfl/, accessed 6 March 2018

Lüse diqiu (Green Earth) http://www.lvsediqiu.com/, accessed 6 March 2018

Paris Agreement http://unfccc.int/paris_agreement/items/9485.php, accessed 6 March 2018

Solidwaste.com.cn http://www.solidwaste.com.cn/, accessed 6 March 2018

Taoqibao vacancies http://web.taoqbao.net/contact_us.html, accessed 6 March 2018

Urban Management and Science and Technology (城市管理与科技) http://www.bjmac.gov.cn/sy/syztzl/M/, accessed 6 March 2018

Wikipedia (Henan Province, Gushi District) https://zh.wikipedia.org/wiki/%E5%9B%BA%E5%A7%8B%E5%8E%BF, accessed 6 March 2018

Zai Shenghuo (A New Living) vacancies http://www.anewlives.com/jionUs01.html, accessed 6 March 2018

Index